AF540578

AN INTRODUCTION TO NUCLEAR CHEMISTRY

AN INTRODUCTION TO NUCLEAR CHEMISTRY

M. SATAKE
Y. MIDO

Consulting Editors
M. S. SETHI
S.A. IQBAL

DISCOVERY PUBLISHING HOUSE
NEW DELHI—110 002

First Published - 1995

Reprinted - 2016

ISBN: 978-81-7141-277-8

An Introduction to Nuclear Chemistry

Published by:

DISCOVERY PUBLISHING HOUSE PVT. LTD.

4383/4B, Ansari Road, Darya Ganj
New Delhi-110 002 (India)
Phone: +91-11-23279245, 43596064-65
Fax: +91-11-23253475
E-mail: discoverypublishinghouse@gmail.com
sales@discoverypublishinggroup.com
web: www.discoverypublishinggroup.com

Printed at:
Infinity Imaging Systems
Delhi

Preface

The teaching of Chemistry at the introductory stage becomes each day a more challenging task as the subject matter becomes more diverse and more complex. These challenges have evoked a series of responses—the present set of introductory chemistry monographs is one such. The teaching of chemistry recognises a number of problems that confront those who select text books. In order to overcome these problems, this volume "**An Introduction to Nuclear Chemistry**"—one of about fifty in the Chemistry Monograph Series—is introduced. Each volume is independent of the others deals with one of Chemistry topics and constitutes and complete entity. Each volume is more comprehensive than can be possible in a single volume text. It is intended to provide a range of topics to cover most undergraduate and chemistry main courses of study. These volumes can be used to enrich the more conventional courses of study.

Suggestions for improvement are welcome and shall be gratefully acknowledged.

Authors

Contents

1

The Development of Nuclear Chemistry

Nuclear Chemistry

To begin with, the term "nuclear chemistry" is a paradox. Chemical changes involve the transfer or sharing of valence electrons. However, the usual properties of nuclei—mass, atomic number, and nuclear structure—bear no direct relation to chemical reactions. Consequently, there can be no direct relationship between the chemical and nuclear properties of atoms.

On the other hand, a wide variety of phenomena are of the interest to chemists either because certain nuclear properties provide new chemical tools or because differences in nuclei actually produce subtleties of chemical behaviour. Thus, nuclear chemistry includes any of the relationship between atomic and nuclear structure of importance to a chemist.

In general, the nuclear chemist is concerned with those properties involving similarities and differences among the isotopes of a given element. It should be stressed that, in addition to stable isotopes, the nuclear chemist studies radioactive isotopes by the application of the various techniques of radio-chemistry and radiation chemistry. We often use the term "radiochemistry" to describe work with radioactive atomic species, whereas radiation chemistry is now limited to chemical effects brought about by radiation interacting with a system.

The development of nuclear science can be divided into two sections. The first covered the period from the discovery of radioactivity until the first World War and provided the information which led to our present concept of the structure of the atom, while the second era, particularly the period between the two World Wars has led more especially to a knowledge of the structure and properties of nucleus.

Stage 1. The Period up to 1914 (*Atomic Structure*)

The first stage in the development of nuclear science begins before the turn of the century wit three basic discoveries, which in chronological order were:

1895—the discovery of X-rays by Roentgen,

1896—the discovery of radioactivity by Becquerel, and

1897—the work of Thomson with cathode rays, culminating in the discovery of the electron and its properties.

These practical discoveries were shortly followed by the development of Einstein's Theory of Relativity, although at the time, its relevance to atomic and nuclear structure was not appreciated.

(A) The Discovery and Properties of the Electron

During the latter half of the last century, a large number of experiments were performed to study the nature of electricity. In one of these, Heinrich Geissler, a German instrument maker, sealed metal electrodes in opposite ends of a glass tube and reduced the air pressure inside to a few mm of mercury. When connected to a source of high potential, a luminous glow was observed inside the tube. This was in fact, the forerunner of the present day "neon" advertising signs.

An Englishman, Sir William Crookes, repeated this work but was bale to obtain a better vacuum. He showed that as the pressure was reduced, the luminous glow broke up and was replaced by a green fluorescence in the walls of the glass tube furthest from the cathode. This he attributed to rays being emitted from the cathode which he named "cathode rays".

At Cambridge in 1897, Sir J.J. Thomson showed that the cathode rays were, in fact, small charged particles which he named "corpuscles"—later to be renamed electrons. He determined the charge to mass ratio by deflecting a narrow stream of these rays with a magnetic field and by observing deflection on a fluorescent screen (Fig. 1.1a). The force exerted by a magnetic field H on a particle of mass m and charge e, moving with a velocity v, is Hev; this is balanced by the centrifugal force of the moving particle acting along the radius (r) of the curved path followed by the particle in the field, *i.e.*

$$Hev = \frac{mv^2}{r}$$

or (1)

$$\frac{e}{m} = \frac{v}{Hr}$$

The velocity *v* is an unknown quantity and must be eliminated from the equation. If an electrostatic field *E* which exerts a force *Ee* on the moving particles is applied simultaneously with the magnetic field, but in a direction so as to return the deflected beam to its original path (*i.e.* an electric field producing a force equal, but opposite, to that resulting from the magnetic field), then:

$$Ee = Hev$$

or (2)

$$v = \frac{E}{H}$$

Substituting (2) in (1) we then have

$$\frac{e}{m} = \frac{E}{H^2 r}$$

Thomson thereby determined an average value for *e/m* of approximately 2×10^{17} e.s.u./g which compares reasonably well with the present value of 5.27×10^{17} e.s.u./g. The corresponding value of *v*, the velocity of the particle, was determined from (2) as 2.8×10^9 cm/sec, or about one-tenth the velocity of light. He further showed that the characteristics of these particles were independent of both the cathode material used, and the chemical nature of the residual gas in the tube. Therefore he concluded that they were of subatomic origin, and suggested that different atoms may, in fact, be different assemblages of these electrically charged particles.

It should be emphasized that the experiments carried out by Thomson led to a fairly accurate determination of the ratio of *e/m*, but not of the absolute value of *e* or *m*. Earlier work on electrolysis by Stoney, had given a rough value of 3×10^{-11} e.s.u. for the electronic charge, and subsequent work by Wilson in 1903 yielded a more accurate value of approximately 10^{-10} e.s.u. It remained for Millikan in 1911 to provide the first accurate determination of electronic charge.

Millikan's apparatus, shown diagrammatically in Fig. 1.1b, consisted of two circular metal plates 22 cm diameter, 1.6 cm apart in a vessel which contained air at a reduced pressure. The upper plate had small holes through which could pass small electrically charged droplets of oil produced by an atomiser. Individual drops which found their way through one of the small holes in the upper plate were observed

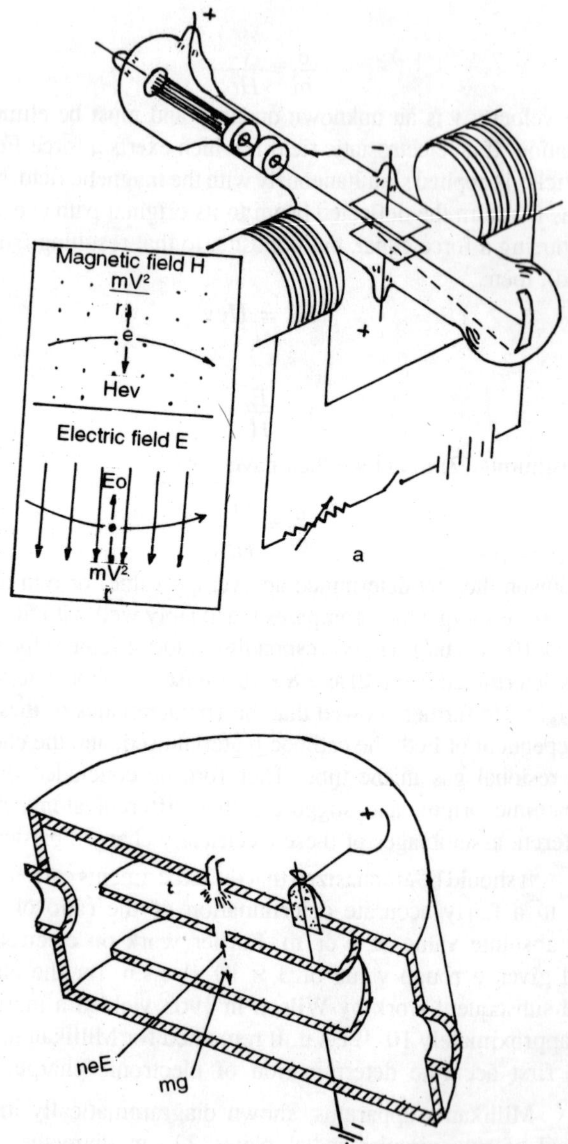

Fig. 1.1 (a) Thomson's apparatus for the determination of the charge to mass ratio of the electron. (b) Millikan's apparatus for the determination of electronic charge.

by a graduated eyepiece in the side of the vessel, and their motion under the influence of a downward gravitational field and upward electrostatic filed was observed. The theory of the method is most easily understood by considering the specific case where the velocity is zero *i.e.* where the downward force on a droplet of mass m is exactly balanced by the upward force of the electric field E.

$mg = neE$ where n is the number of electronic charges e on a drop.

$$ne = \frac{mg}{E}$$

The value of m was determined from the volume of the drop, which in turn, was obtained by observing the rate of fall in the absence of the electric field, and applying an equation derived by G. G. Stokes

$$v = \frac{2gr^2d}{9\eta}$$

where v is the terminal velocity due to gravity,
g gravitational acceleration,
r the drop radius,
d is the density of the oil,
η is the coef, of viscosity of air.

A very large number of determinations showed ne to be an integral multiple of a single value, the electronic charge, to which Millikan ascribed a value of 4.77×10^{-10} e.s.u., very close to the present accepted value of 4.803×10^{-10} e.s.u. (1.60×10^{-19} coulombs).

(B) X-rays

Whilst experimenting with cathode rays, the German physicist Roentgen found that some chemicals would fluoresce some distance away from the cathode ray tube. Experiments showed that these rays, which had considerable range and penetrating properties, originated from the region where the cathode rays impinged upon the anode end of the tube. He called them "X-rays" and announced his discovery to the German Physical Society on Christmas Eve, 1895.

In recognition of this discovery, the unit of radiation intensity has been named after him:

One roentgen (1 r) is that quantity of X-rays or γ-rays which will produce, as a consequence of ionisation, one e.s.u. of electric charge of either sign, in one cc of dry air measured at S.T.P. (i.e. in 1,293 mg of air).

X-rays, which are short wavelength electromagnetic radiation, are produced by the abrupt deceleration of last moving electrons in a target material. An X-ray tube for the production of X-rays has three essential features: a source of electrons (invariably a heated filament), a suitable high accelerating potential—usually of the order of 50,000 volts, and an anode target constructed from a metal of high atomic number.

The X-rays emitted have two distinct components:

(*i*) X-rays with wavelengths distributed throughout the X-ray spectrum, similar to the continuous distribution of wavelengths in white light. These X-rays result from the slowing down of electrons as they pass close to the nuclei of atoms, and they may be compared with bremsstrahlung radiation, which is produced by the deceleration of β-particles in the vicinity of nuclei.

(*ii*) X-rays with characteristic fixed wavelengths superimposed on the continuously distributed wavelengths mentioned above. These X-rays are produced when one of the bombarding electrons ejects and inner bound orbital electron, and an outer electron drops into the vacant position. In so doing, it releases an X-ray the wavelength of which depends upon the energy of the atomic orbital being filled. These X-rays may be identified with the X-rays emitted after certain modes of radioactive decay *e.g.* electron capture.

(C) Radioactivity

The effect which uranium salts could produce on a photographic plate was reported as early as 1867 by N. de St. Victor. The discovery of radioactivity must, however, be accredited to Henri Becquerel. In 1896, during a study of fluorescence in uranium nitrate, he showed that invisible radiations emitted by uranium affected a photographic plate. Furthermore they were independent of any external conditions applied to the uranium nitrate and therefore were not a fluorescence phenomenon. The discovery was one more of accident than design, and occurred in the course of studying the fluorescent properties of various materials. It is interesting to speculate how long radioactivity would have remained undiscovered, and what our present knowledge would be, if it were not for the pure coincidence that the nitrate of uranium is one of the few inorganic compounds which fluoresce; a property which has no connection with its radioactive properties.

It was perhaps fortunate that the Curies were working in Becquerel's laboratory at this time and undertook a systematic study of this new phenomenon. They provided what they considered existence for two new elements, polonium and radium, which in the course of the next few years they succeeded in isolating in weighable quantities; they also began a systematic study of the emitted radiations.

In honour of the Curies' work, the unit of radioactivity was named the curie. Originally it was defined in terms of radon activity in equilibrium with 1 g of radium—which was from time to time subject to revision. It was finally defined in 1948 in terms of an absolute value namely:

One curie is that quantity of radioactive material giving 3.7 × 10^{10} disintegrations per sec (i.e. 2.2 × 10^{12} disintegrations per min.).

As a result of the work of Rutherford in England, and the Curies in France, the existence of three different types of radiation with different characteristics of range, penetration and charge was established by 1903. They were designated by the first three letters of the Greek alphabet, α, β and γ. In the following ten years, Rutherford, with a succession of coworkers, many of whom were to reach world fame, made an exhaustive study of the positive α-particles which were emitted. Much of Rutherford's early work was carried out at Montreal* in conjunction with the chemist F. Soddy. Rutherford subsequently moved to Manchester and finally back to Cambridge, where he had begun his work as a student of J.J. Thomson.

In 1895, just before the discovery of radioactivity, Sir William Ramsay isolated the inert** gas helium, and showed that it occurred in minerals which contained either uranium or thorium. Following the discovery of radioactivity, Rutherford and Soddy suggested that helium was one of the ultimate products of radioactive decay, and therefore accumulated in minerals containing these radioactive elements. In 1903 Ramsay and Soddy condensed a small amount of radon (the radioactive, but chemically inert gas produced in the radioactive decay of radium) in a tube cooled by liquid air. After pumping off all non-condensable gases, they sealed the radon in a small spectrum tube. Initially no helium was detected, but over a period of several days, the helium spectrum

* One of Rutherford's very early co-workers at Montreal was O. Hahn, who subsequently discovered nuclear fission.

** The "inert" gases are here so named for historical reasons. It was shown in 1962 that under certain conditions they can in fact form chemical compounds. The name "*noble gases*" is now preferred.

began to appear, increasing in intensity as time went on. It was subsequently shown that helium was evolved by several radioactive substances, all of which have one thing in common—they all decayed by α-particle emission.

In 1909 Rutherford and Royds established the identity of α-particles as helium nuclei. They sealed radon in a thin walled glass capillary tube, which was contained within a larger tube. The wall of the capillary was thin enough to allow α–particles to pass through and thereby to accumulate in the space between the two tubes. After a period of several days, the gas present in the outer tube was identified spectrographically as helium. In scattering experiments with α–particles—carried out between 1909 and 1911 at Manchester with Marsden and Geiger—Rutherford obtained evidence of the dimensions and the nature of atomic nuclei which led directly to our present picture of atomic structure. When thin metallic foils were bombarded with α–particles, some of the α–particles passed straight through showing that metals were not composed of closely packed impenetrable atoms but were largely empty space. Other α–particles were deflected through very great angles from their original path. This pointed to the existence of localized, but very strong, positive electric charges (Fig. 1.2a)

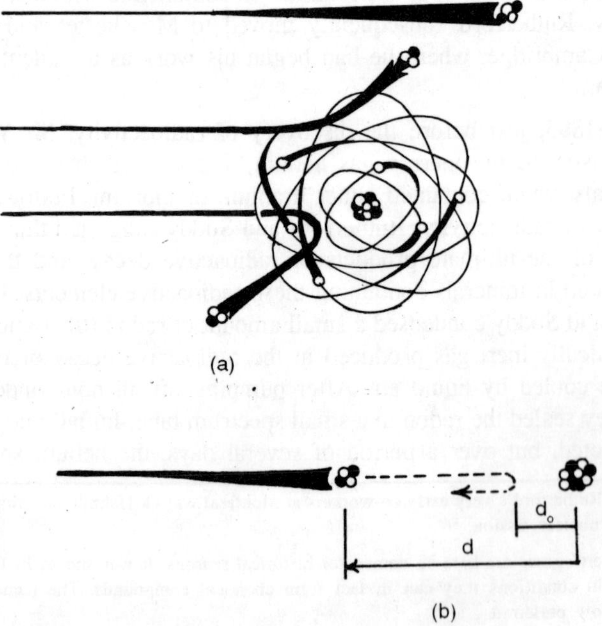

Fig. 1.2. Deflection of α-particles by atomic nuclei.

A careful study of the angular distribution of the deflected particles, showed that the scattering was due to a single head-on encounter with an atomic nucleus, and it was possible to use the results to estimate the dimensions of the nucleus. The general underlying principle can be very simply illustrated by considering the specific case of closest approach, when an α–particle is reflected back along its original path (Fig. 1.2b).

From electrostatics:

Force of repulsion between target nucleus and α–particle is:

$$\frac{2e \cdot Ze}{d^2}$$

where e is electronic large,
Z is atomic No.,
d is distance apart.

and potential energy (or work done by the repulsive force in bringing the particle to rest) is

$$\frac{2Ze^2}{d}$$

The α-particle will come to rest and be started back along its track when its initial K.E. = P.E. from repulsion,
i.e. when

$$^1/_2\, mv^2 = \frac{2Ze^2}{d_0}$$

or

$$d_0 = \frac{4Ze^2}{mv^2}$$

where d_0 is the distance of closest approach.
Considering a typical case of a 5 MeV α-particle approaching an atom of Al (Z = 13), the substituting 5 MeV (= $5 \times 1.6 \times 10^{-6}$ ergs) for ½ mv^2 we have

$$d_0 = \frac{2 \times 13 \times (4.8 \times 10^{-10})^2}{5 \times 1.6 \times 10^{-6}} = 7.5 \times 10^{-13} \text{ cm.}$$

These observations led Rutherford to revise and elaborate on an earlier atomic model suggested by Nagaoka in 1904. Rutherford's model consisted of a heavy positively charged central nucleus with a diameter of between 10^{-13} and 10^{-12} cm about which electrons revolve in orbits with a radius of the order of 10^{-8} cm.

This early physical picture of the atom was further developed in 1913 by Niels Bohr, while working in Rutherford's Manchester Laboratory, and placed on a more quantitative basis with the postulate of closed electron orbits or "orbitals" which, contrary to classical theory, represented "stationary" states in which the electrons could move without radiating energy. In this same year Rutherford identified the proton (the hydrogen ion) as the positive counterpart of the electron, and Moseley working in Rutherford's laboratory at Manchester made a systematic study of X-rays emitted by various elements. From this work he concluded that the number of charges in the nucleus increased by one electronic unit from atom to atom, in the sequence of the elements into the Periodic Table.

Briefly, the picture of the atom which has subsequently evolved is that of a positively charged nucleus containing 99.98% of the mass of the atom, about which electrons are arranged in energy levels. These are described in terms of four quantum numbers:

(*i*) *The principal quantum number (n)* which may take values, 1, 2, 3, etc. and which defines the electron shell being considered. The innermost shell $n = 1$ is sometimes referred to as the K shell the next ($n = 2$) as the L shell and so on.

(*ii*) The azimuthal quantum number (l) which may take values, O, 1, 2, 3, etc. and designated as *s, p, d, f* with a restriction on an upper limit given by $l \ngtr (n-1)$. This quantum number which is usually associated with orbital shape is used in combination with the principal quantum number to define the subshells e.g. 1*s*, 2*s*, 2*p*, etc.

(*iii*) *The magnetic quantum number (m)* associated with spectral behaviour in a magnetic field and which may take integral values from $-l$ to $+l$ thereby providing $(2l+1)$ subdivisions, or multiplicity, of the energy levels for each value of l.

(*iv*) *The spin quantum number (s),* associated with the "spin" of the electron, and which may take values of either $+\frac{1}{2}$ or $-\frac{1}{2}$, thereby providing a total of $2(2l+1)$ subdivisions for each value of l.

Pauli's exclusion principle which requires that no two electrons may occupy the same energy state (*i.e.* have all four quantum numbers the same) permits us to determine the number of electrons in any one shell or subshell. For the *s* subshell the number is 2, for the *p* subshell 6, for the *d* subshell 10 and for the *f* subshell 14. The subshell fill in the order of their energy, which is *not* in a strict numerical quantum number sequence. The stable configurations which correspond to the stable inert gas structures correspond to regions where there is a large energy gap between shells being successively filled. The sequence whereby subshells are filled leads to the familiar 2,8,8,18,18,32 pattern.

Electrons may be promoted from one energy level to another, or completely removed from the atom to form an ion, by the application of energy, either chemical—from a chemical reaction—or physical *i.e.* by the absorption of electromagnetic radiation. Less tightly bound electrons in the outer energy level can be excited to a higher level by electromagnetic radiation in the visible and near ultraviolet spectrum, equivalent to only a few electron volts. Frequently the absorption of such energy is immediately followed by the emission of light, as the electron drops back again—this is the phenomenon of *fluorescence.* If in order to get back, the electron finds itself in a state where it must reverse its spin, as is often the case in certain molecular structures, the transition is said to be "forbidden" and the process is consequently delayed for a measurable period of time. The phenomenon is then referred to as *phosphorescence.* High energy processes, of the order of thousands of electron volts per atom, such as high energy electron bombardments, γ-radiation or some nuclear processes, lead to the removal of an electron from an inner atomic shell *i.e.* the *K* or *L* shell. This is followed by the refilling of the level accompanied by the emission of an X-ray of energy characteristic of the vacant electron level, and frequently referred to as *K* or *L* X-rays depending on the atomic level concerned. As is well known, our present concept of the atom includes the notion that the atom consists of a dense nucleus containing protons and neutrons and surrounded by electrons occupying certain positions, or following certain paths of motion. It would be very simple if we could describe the electrons as having definite kinds of motion. However, it was pointed out by Heisenberg in his statement of the uncertainty principle that it is impossible to know at the same time the energy and the position of any particle such as an electron. Any experiment designed to measure the location or the energy of an electron must change either the location or the energy in the measuring process. *Since we cannot specify the electron motion precisely, it is impossible to draw electron paths or tracks. Hence, we can speak only of the probability or relative chance of finding an electron at any given position within the atom.*

It is important to realize that a proper description of atomic or subatomic structure can be made only through the use of mathematical equations. Even so, it is worthwhile and quite common to employ what the scientist call a "model" for descriptive purposes. The particular model he will use varies with his sophistication. The development of scientific models that are incomplete, if necessary, but still not incorrect is one of the interesting problems of scientific communications.

Finding an Electron by Quantum Mechanics. The probability of finding an electron in an atom can be deduced by means of the techniques of quantum or wave mechanics. The common method of using these mathematical techniques is to formulate the equations that describe the motions of waves and to employ them to describe the probabilities of finding atomic or subatomic particles. One of the results of describing the probability of finding an electron is shown in Fig 1.3. This diagram illustrates the probability of finding the electron nearest the nucleus at a given point in the atom of which it is a part. The probability of locating it in a particular small region (at a given point) is shown, and it can be seen that the probability of finding this electron is highest at the center of the nucleus. The motion corresponding to this probability can be described as an oscillating motion through the nucleus, but as we shall see later, an electron cannot be considered as a part of the nucleus.

On the other hand, we can show the probability of finding the electron in a particular spherical "shell" outside of the nucleus. The curve is shown in Fig. 1.4. The difference in these two diagrams is

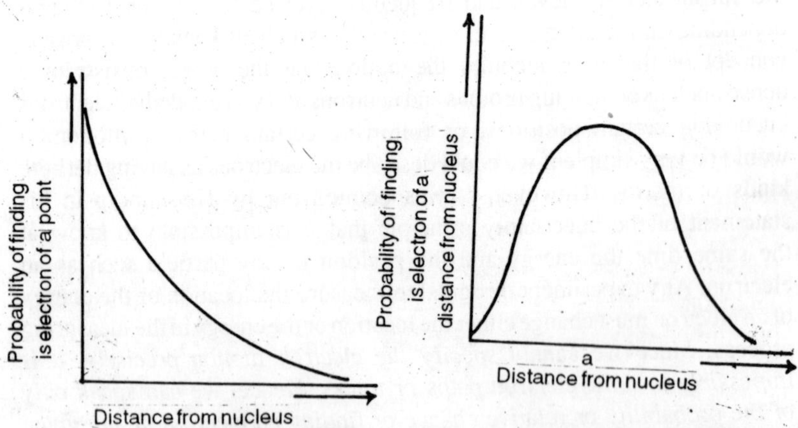

Fig 1.3. Probability of finding a ls electron of a given point in the atom.

Fig 1.4. Probability of finding a ls electron in a given shell of the atom.

explained on the basis of the fact that although the probability of finding the electron in a given small volume of the atom is highest at the center, and decreases as one goes away from the nucleus, the number of these small volumes in each shell around the nucleus increases as the shell becomes larger.

Principles of Quantum Mechanics. The basic ideas of atomic structure were stated by Bohr some years ago and have been retained in essentially the same form in our present atomic theory. This theory has been of striking consequence in explaining a wide variety of experimental results. The various probability distributions that are possible correspond to different energy states of the electrons, and these energy-level designations are used to describe the electrons found in the various atomic shells. The innermost energy level is referred to as the *K* level, the next level as the *L*, level, and so forth. These principal levels are further described as having various sub levels denoted by the letters *s, p, d,* or *f*.

One of the basic principles of quantum mechanics is that only specified energy levels are possible for electrons in atoms. These levels are numbered with the *K* level having the principal quantum number, designated $n = 1$, the *L* level having $n = 2$. A second principle is that the electron population of any energy level in an atom is limited to $2n^2$. This means that for the *K* level, the maximum number of electrons permitted is $2\ (1)^2$ or 2; for the second level, the maximum number is $2\ (2)^2$ or 8. Another postulation of the theory is that each main level of electrons may have a number of sub levels equal to its principal quantum number. This would give one sublevel to the *K* shell, two subshells to the *L* level, with the *M* level having three. Just as the number of electrons that can be put in any main shell is limited, so is the number of electrons in each subshell. An *s* subshell can hold two electrons, a *p* subshell six, a *d* subshell ten, and an *f* subshell fourteen. It is well known from chemical valence values that the outer shells do not contain more than eight electrons each.

The Periodic Table. From this kind of outline it is possible to describe all of the atoms of the elements of the periodic table. It might seem that each of the levels and sublevels should be filled in order, but this is not always the case. For example, it is found that the 4*s* level has lower energy than the 3*d* level. This means that, in some elements, electrons would be found in the 4*s* state before the 3*d* state is filled. Table 1.1 shows the electronic configurations of a few elements. The superscripts on the sublevel letters indicate the number of electrons in that particular subshell.

Table 1.1 : Electron Populations of Levels and Sublevels

Atomic Number	*Element*		*Usual Chemical Valence*
1	H	$1s^1$	1
2	He	$1s^2$	0
3	Li	$1s^22s^1$	1
4	Be	$1s^22s^2$	2
10	Ne	$1s^22s^22p^6$	0
11	Na	$1s^22s^22p^63s^1$	1
21	Sc	$1s^22s^22p^63s^23p^63d^14s^2$	3
57	La	$1s^22s^22p^63s^23p^63d^{10}4s^24p^64d^{10}5s^25p^65d^1 6s^2$	3

The information about these electronic configuration is obtained primarily from spectroscopic data in which the various spectral lines are considered as arising as a result of transitions of electrons from one energy level to another. Of course the chemical behaviour of the elements is determined by the distribution of electrons in the atom with particular significance being attached to the number and arrangements of the outermost or valence electrons. *One of the paradoxes involved in the term "nuclear chemistry" is the lack of a relationship between the nuclear properties of an atom and its chemical behaviour.*

Electron Clouds and Energy Level Diagrams. It was pointed out that we cannot describe the motions of the electrons in the atom even though we can speak of their energy levels and can give accurate descriptions of atoms by describing the number of electrons in the various shells and subshells. Since we often think in three dimensional analogies, let us suppose that an atom were large enough to be seen and that a trail of smoke were left by each electron in the atom. The density of the smoke would tell us where the electrons had been, and from this the probability of finding the various electrons at any point in the atom could be determined. In Fig. 1.5a, the electron cloud is represented as a three-dimensional ball of smoke with no well-determined pattern of motion or structure.

If, however, this electron cloud is cut in cross section (Fig 1.5b), we should be able to see a series of regions in which the electron cloud is more dense. Although this distinctness of structure would indicate that the probability of finding the electrons in certain regions of the atom is higher than finding them in other regions, there still would be no manifestation of precise motions. Even though the modern picture does not describe the atom as containing electrons in "orbits," we can determine the location of the highest electron probability which does correspond to the radius of the orbit as described in the earlier static models of the atom. (For convenience, sometimes chemists still draw such plane orbits, even though they visualize the electrons as occurring only in clouds.) Our concern is, thus, not so much with the positions of the electrons as with their energies. We may point out that an electron belonging to a given level of any atom may be found anywhere in space, but it does not use up its own energy in passing from one point to another so long as it is still associated with that atom. Hence, it is not necessary to draw circles to represent these electron energy levels. The same idea could be conveyed by drawing horizontal lines (Fig 1.5c) to represent the electron levels. Such an "energy level diagram" is useful as a way of visualizing energy relationships, and is used extensively in modern description of atomic and subatomic phenomena.

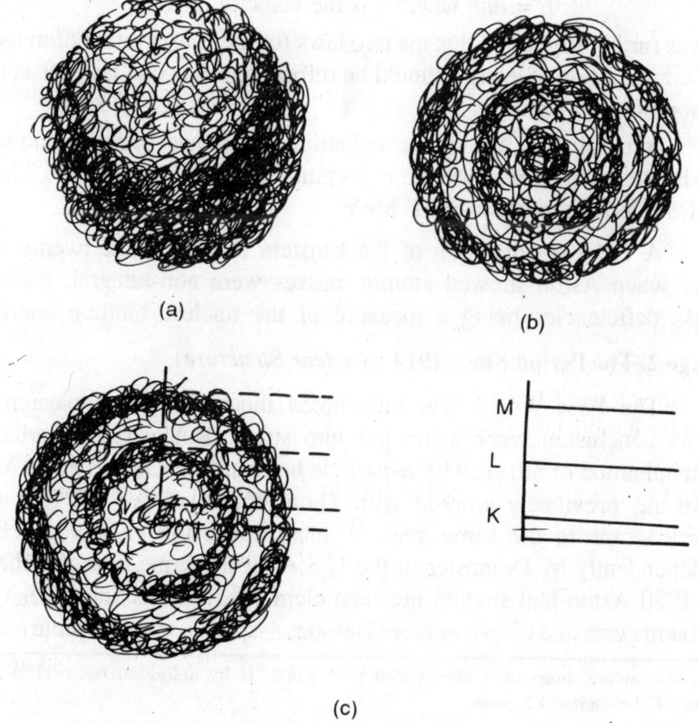

Fig. 1.5 : Electron Cloud Density and Energy Levels for Electrons in an Atom

The foregoing summary of the extranuclear atom has been included, not for the usual purpose of introducing the chemical properties of the elements, but because a somewhat similar line of reasoning is employed in the nuclear shell theory of nuclear structure, to be discussed later and also to revise the concepts of fluorescence, phosphorescence and X-ray emission, which will be encountered in subsequent chapters.

(D) Einstein's Special Theory of Relativity

1905 saw a very important milestone in nuclear science in the form of Einstein' Special Theory of Relativity, although at the time, its relevance to nuclear science was not fully appreciated. In the course of developing a new system of mechanics based upon the hypothesis that the velocity of light was a constant independent of the relative motion of the source and observer, Einstein showed that there existed an equivalence between mass m and energy E given by:

$E = mc^2$ where c is the velocity of light.

It was further suggested that the two laws for the conservation of mass and the conservation of energy should be reformulated as one law, the conservation of mass-energy.

According to the above equation which has become known as the Einstein equation, 1 atomic mass unit* (1.66×10^{-24} g) = 931 MeV** and 1 electron mass = 0.51 MeV.

A fuller appreciation of the Einstein relation came twenty years later when Aston showed atomic masses were non-integral, the slight mass deficiencies being a measure of the nuclear binding energy.

Stage 2. The Period Since 1914 (*Nuclear Structure*)

The First World War interrupted much scientific research, but at its conclusion work again got into stride. Rutherford reported the transmutation of nitrogen by α-particle bombardment in 1919 and Aston, who has previously worked with Thomson, developed his first mass spectro*graph* in the same year. A mass spectro*meter* was developed independently by Dempster in the U.S.A. at the same time. By the end of 1920 Aston had studied nineteen elements, and has shown that nine of them consisted of two or more isotopes—species with the same nuclear

---

* One atomic mass unit, abbreviated to 1 a.m.u. is by definition one-twelfth of the mass of the carbon-12 atom.

** MeV (million electron volts) is the energy unit normally used in this subject. It is equal to the energy which would be acquired by an electron accelerated through a potential difference of 10^6 volts and is equivalent to $1.6 \cdot 10^{-6}$ ergs.

charge, but different nuclear mass. He thereby paved the way for Rutherford to postulate the existence of the neutron, which was not to receive experimental confirmation for another twelve years.

The twenties did not represent a period of great new discoveries but rather a period of refinement and improvement of techniques and equipment, and further investigation of discovered phenomena. Two important discoveries, however, were made by Aston with his improved high resolution mass spectrograph. With his new instrument, he showed that nearly all natural elements existed in isotopic forms and that atomic masses were in fact, not exactly whole numbers. This latter discovery along with Einstein's $E= mc^2$ relation (1905) opened the way to an understanding of nuclear binding energy, and of the energy which was available from nuclear sources.

The thirties began with a rush of new discoveries. Four artificial accelerators—the Van de Graaff generator, the Cockroft—Walton voltage multiplier, the linear accelerator, and the cyclotron were all developed in the first few years of the decade. Cockroft and Walton working in Rutherford's Cavendish laboratory, were the first to produce transmutation by purely artificial means. In the same year, 1932, and in the same laboratory Chadwick discovered the neutron following a decade of work begun under the direction of Rutherford. This represents one of the most important milestones in nuclear science.

In actual fact, the first experimental evidence for the existence of the neutron was supplied by Bothe and Becker in 1930, when they found that α-particles produced an unusually penetrating radiation when they fell on certain light elements. The details of the experimental results were difficult to explain on the basis that the radiation was γ-radiation, and the following year the daughter of Mme. Curie, Irene Curie, with her husband F. Joliot, showed that this radiation was capable of ejecting protons of high energy from paraffin. It was left to Chadwick to interpret these results correctly and to add additional experimental evidence to verify the existence of the neutron, which had been postulated by Rutherford in his Bakerian Lecture to the Royal Society in 1920.

The year 1932 also saw the identification of the positron (*i.e.* the positive electron) by Anderson in Berkeley, California.

The remainder of the thirties was notable mainly for the investigation of nuclear reactions using artificially accelerated particles, and for an extensive study, particularly by Fermi, of neutron induced reactions which culminated in the discovery of fission by Hahn and Strassmann at the end of 1938.

The discovery that uranium nuclei split up when bombarded by neutrons was not as important as the fact that the event was accompanied by the emission of more neutrons than were necessary to produce fission, thus providing evidence for the feasibility of a chain reaction. What is not generally known about the discovery of fission is that evidence for the phenomenon was recorded as far back as 1934 by Fermi, in a paper entitled *Possible Production of Elements of Atomic Number Higher than 92.* However, in this paper, Fermi mistakenly claimed to have produced unidentified elements higher than uranium instead of making a more critical appraisal of what was evidence. of fission. From the neutron irradiation of uranium, active isotopes with half-lives "of about 10 sec, 40 sec, 13 min, plus at least two more periods from 40 minutes to one day are well established" claimed Fermi. The fact that so many different radioactive species were apparently present did not lead him to look for possible neutron reactions other than those already established, Despite all the half lives claimed, only one—the 13 minute activity — was subjected to any chemical tests for its identity in Fermi's 1934 paper.

Failing to identify the 13 minute activity with uranium or any of the known elements immediately below uranium, Fermi incorrectly supposed that the activity was due to an element of atomic number higher than uranium. Fermi, after the discovery of fission, is recorded by his wife in his biography "*Atoms in the Family*" as saying : "We did not have enough imagination to think that a different process of disintegration might occur in uranium from that in any other element, and we tried to identify the radioactive products with elements close to uranium in the Periodic Table of Elements. Moreover, we did not know enough chemistry to separate the products of uranium disintegration from one another"*.

It is also interesting to speculate on the effect which the proper understanding of fission by Fermi in 1934 would have had on the outcome of the Second World War.

It was left for two chemists, Hahn and Strassmann, working at the Kaiser Wilhelm Institute in Berlin to identify an isotope of barium

* Fermi must, however, have been aware of the Noddack's paper (*Z. Angew. Chem.*, **47**, 633, 1934) in which a new mode of disintegration was suggested as an alternative explanation for Fermi's results.

as one of the products obtained by irradiating uranium with neutrons, and to announce the discovery of fission to the world in a paper dated 22nd December 1938 and published in *Naturwissenschaften* in January, 1939. It is perhaps appropriate to point out that this discovery—which opened the door to the exploitation of nuclear energy—was made only fourteen months after the death of Lord Rutherford.

At the present time, when nuclear weapons and nuclear power are taken for granted, it is interesting to look back to 1920. The electron and proton were the only particles then discovered. The Bohr—Rutherford model of the atom had just been accepted and the scientific world was still marvelling at Aston's first—though somewhat crude—mass spectrograph. In that state of knowledge and development it is interesting to look at the prophetic remarks of an early pioneer in the field, Frederick Soddy, who in his book The *Interpretation of Radium* (1920) wrote: "The problem of transmutation and the liberation of atomic energy to carry on the labour of the world is no longer surrounded with mystery and ignorance, but is daily being reduced to a form capable of exact quantitative reasoning. It may be that it will remain forever unsolved. But we are advancing along the only road likely to bring success at a rate which makes it probable that one day will see its achievement.

Should that day ever arrive, let no one be blind to the magnitude of the issue at stake, or suppose that such an acquisition to the physical resources of humanity can safely be entrusted to those who in the past have converted 'the blessing already conferred by science into a curse".

2

Fundamental Particles and Nuclear Structure

The discovery of the first fundamental particles, the electron, the proton and the neutron, together with the formulation of the concept that atoms have a structure and that different atomic species's owe their different properties to different arrangements of these few fundamental particles, brought about an amazing simplification and understanding of chemical theory. Just three different building's blocks to worry about instead of the 90 or so different chemical elements. All would have been well—for the chemist at least—if things had stopped there, but they didn't. As the years have rolled by, more and more particles have been discovered, until the stage has now been reached where there are a third as many particles as there are chemical elements. It is, therefore, not surprising that some physicists are speculating on the possibility that the "fundamental" particles may in fact have some sort of internal structure—or that some particles may physicists are speculating on the possibility that the "fundamental" particles may in fact have some sort of internal structure—or that some particles may be excited states of others. There is certainly experimental evidence to suggest that the charge distribution in protons, for example, is not uniform. While these points should be borne in mind, for the purpose of this book we shall consider the "fundamental" particles as truly fundamental.

Fundamental Particles

Besides the γ-photon, 32 fundamental particles have been postulated, and experimental verification of the identity of all but a few has been achieved. Fortunately, a knowledge of only a small fraction of these is necessary in understanding atomic structure and nuclear properties. The important properties by which these particles are defined are their charge, mass and spin.

(A) The Electron

The first fundamental particle to be discovered—the electron—is probably the most important. It is the particle responsible for the formation of chemical bonds, and for the passage of an electric current through a conductor. It appears as cathode rays in a cathode ray tube and is the β-particle emitted during radioactive decay. As mentioned in the previous chapter, the electron carries a charge of 4.8×10^{-10} e.s.u. (1.6×10^{-19} coulombs) and has a mass of 0.00055 atomic mass units (9.108×10^{-28} g).

(B) The Proton

Besides work on the cathode rays, several investigators looked at the properties on the positive or "canal rays" which could be produced by an electric discharge in gases at low pressure. The lightest such particle to be discovered was that due to ionised hydrogen which, in 1914, Rutherford suggested was the long sought after positive counterpart to the electron. At Rutherford's suggestion, the term proton was adopted as the name for this particle at the 1920 Cardiff meeting of the British Association.

Thus the proton, with its electric charge equal in magnitude but opposite in sign to the electron, and with a mass some 1800 times that of the electron (1.00728 a.m.u.) was recognised as one of the fundamental building blocks of atoms.

(C) The Neutron

The development of the mass spectrograph and its use by Aston in 1919 provided evidence for the existence of isotopes of nine elements, i.e. atoms with identical chemical properties and hence the same number of electrons and protons, but with different masses. This led Rutherford to postulate the possible existence of a neutral particle with a mass similar to that of a proton, during the course of the Bakerian Lecture to the Royal Society in 1920. In 1921 Harkins applied the name neutron to this, as yet, undiscovered particle.

Because of the absence of any electric charge and its consequent inability to produce ionisation by other than indirect means, the existence of this particle was not established for over a decade. It had been shown by Bothe and Becker in 1930 that α-particle bombardment of some light elements, notably beryllium, produced radiation even more penetrating than γ-rays. Further, it was shown by Mme. Joliot–Curie in 1931, that if paraffin wax was placed between the source of the radiation and an ionisation detector, protons were ejected from the wax. Chadwick repeated these experiments, and in 1932 published a correct interpretation of the phenomena—the ejection of a neutron from beryllium by α bombardment according to the reaction :

$$^{9}_{4}\text{Be} + {}^{4}_{2}\text{He} \rightarrow {}^{12}_{6}\text{C} + {}^{1}_{0}\text{n}$$

(where, by convention, the superscripts define the mass number and the subscripts define the atomic number of the nuclear species).

In the pre-neutron era, the nucleus was considered to be composed of protons equal in number to the atomic number, together with additional protons and an equal number of electrons, to make up the extra mass with no additional charge.

Chadwick, who discovered the neutron, estimated that the mass of the neutron was just a little less than the sum of the masses of the proton and electron, and wrote : "this suggests that the neutron consists of a proton and an electron in close combination". The fact that a neutron is converted to a proton and an electron in the process of β decay would also support this view were it not for the fact that a proton is converted to a neutron and a positron in the process of β^{+} decay, thus suggesting that the proton was a composite particle composed of a neutron and positron. The most decisive evidence for the neutron as a fundamental particle comes from nuclear magnetic resonance studies, from which a measurement of nuclear spin can be made. Briefly, nuclear spin depends upon the net effect of the particles from which the nucleus is believed to be composed. The hydrogen nucleus (a single proton) has a spin of ½ and the deuteron a spin of 1. Had the deuterium nucleus been composed of two protons and one electron, the overall spin from the odd number of particles would be fractional. Direct measurements of the neutron spin have shown that it is ½ and not 0, or 1, as would be the case if it were composed of two particles.

The present accepted neutron mass is 1.008665 a.m.u.

The existence of the three basic atomic building blocks was thus established to give us our presently accepted picture of the atom as composed of a nucleus containing protons and neutrons, which are frequently grouped under a single term "nucleons" and exterior to which electrons exist in energy levels:

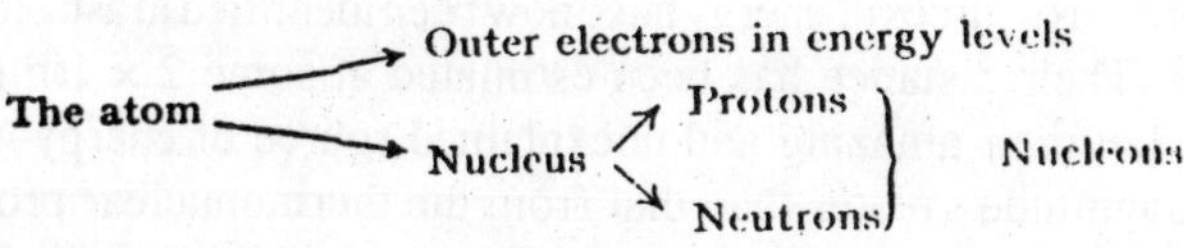

Z (atomic number) = number of protons (also = number of electrons)
N = number of neutrons
A (mass number) = *Z* + *N*

(D) The Positron (e^+)

With the discovery of the neutron in 1932 to complete the three subatomic building blocks, all seemed well, and it was therefore somewhat disconcerting when, at the end of 1932, Anderson found evidence in cloud chamber photographs for a particle very similar to the electron, but of opposite charge. This particle he named the positron. The positron, which is frequently encountered, as a mode of radioactive decay, is now regarded as an "antiparticle". One of the most important characteristics of antiparticle is that when an antiparticle, and the corresponding particle, in this case the electron, come together they mutually annihilate each other, and energy equal to their combined mass is liberated. In the case of the positron and electron, this is in the form of two 0.51 MeV γ-rays. This phenomenon represents an example of the complete conversion of mass into energy. It is also possibly for γ-ray energy to be converted into a positron-electron pair.

Although not essential in the present study of nuclear chemistry, it is of interest to point out that the existence of the antiproton—negatively charged proton—and also the anti-neutron, have been postulated, and their presence confirmed experimentally in recent years using high energy accelerators.

One of the more interesting points of speculation at the present time is the possible existence of anitmatter—composed of antiprotons, antineutrons and positrons—elsewhere in the universe. In all experimentally observed cases, where energy has been converted into matter, particle—antiparticle pairs. If these "antimatter" particles are the bases for other material systems outside our galaxy—and there is no experimental evidence to suppose the contrary—then it is possible the universe consists equally of matter and antimatter. If this were so, however, would we not expect to see some evidence of annihilation processes?

The recently discovered quasars may provide an answer. Several of these enormous sources of energy have now been identified in astronomical observation. Their distance has been estimated at some 2×10^9 to 10^{10} light years, but their amazing and unexplained source of energy—many orders of magnitude greater than that from the thermonuclear processes of the other galaxies—remains a mystery. Besides looking at them through a truly astronomical distance, we are also looking back in time, to a period which is, in fact, close to the estimated age of our own solar system. Are we, therefore, observing an antimatter-matter annihilation process producing, by a spontaneous reaction, the energy which will ultimately "ignite" a thermonuclear reaction among the surplus matter or anitmatter? In short, are we witnessing the birth of a galaxy which at the present time may be, in fact, similar to our own?

(E) The Neutrino (ν)

It had long been known that the energies of the β-particles emitted from the same radioactive atomic species are continuously distributed over a wide range, and not of the same fixed value. In order to explain this and also to conserve "spin" in nuclear decay mechanisms, Pauli suggested in 1931, that a second undetected particle was simultaneously emitted to carry away additional kinetic energy so that the total disintegration energy is shared between the two particles. This idea was developed further by Fermi in 1934, and the particle was named neutrino. Its properties were that it had zero charge and almost zero mass. The corresponding counter-part, the antineutrino, should also exist. In fact, it is the anti-neutrino which is associated with β decay and the neutrino which is associated with positron emission. For the purposes of this book no further distinction will be made between neutrinos and antineutrinos. The existence of the neutrino was experimentally confirmed in 1956.

(F) Mesons

In an attempt to explain nuclear forces, Yukawa in 1934 postulated the existence of yet another particle with a mass intermediate between that of the electron and that of the proton. In 1937 experimental evidence from cosmic ray studies was obtained by Anderson (the discoverer of the positron) for the existence of both positive and negative particles with a mass 207 times that of the electron. These were the μ mesons or "muons" and the problem appeared solved.

However, a study of these particles showed inconsistencies with the properties of Yukawa's postulated particle. Since that time, many other particles belonging to the "meson" family have been discovered—among them the π mesons, or "pions", discovered in 1947 by Powell with a mass of approximately 273 electron masses, and which have the properties required by Yukawa's theory.

Elementary Particles—Summary

Particle	*Symbol*	*Charge (referred to electron as —1)*	*Mass (relative to electron*	*Spin*
Electron (or negatron)	e, β ($e^-\beta^-$)	—1	1	½
Positron	$e^+\beta^+$	+1	1	½
Proton	P	+1	1836.1	½
Neutron	n	0	1838.6	½
Neutrino	ν	0	~0	½
Antineutrino	ν	0	~0	½
μ Mesons (muons)	μ^+	+1	206.9	½
	μ	1	206.9	½
π Mesons(pions)	π^+	+1	273.2	0
	π^0	0	264	0
	π^-	—1	273.2	0

Nuclear Magnetic Moments—Nuclear Magnetic Resonance

Associated with charged particles which have angular momentum is a property called the *magnetic moment*. The unit is the magneton, equal to $eh/4\pi mc$, where e is the charge, h is Planck's constant, c the velocity of light and m the mass of the particle. The *Bohr magneton* is the unit when e is the electronic charge, and m the mass of the electron. The *nuclear magneton* is the unit when m is the mass of the proton and, in theory, this should have a magnitude some 1/1800 that of the Bohr magneton.

If a sample of water (or other hydrogen compound) surrounded by a coil is placed in a strong magnetic field, it is found that at a certain radiofrequency energy is absorbed from the coil. For a field of 10,000 gauss this frequency is 42.6 Mc/s. If this same experiment is repeated, but using heavy water (or other deuterium compound) the frequency shifts to only 6.5 Mc/s. It is evident that this phenomenon is to be associated with the nuclear composition of the two hydrogen isotopes. Other nuclear species also show characteristic resonant frequencies. The phenomenon is attributed to nuclear magnetic moments arising from "nuclear spin". Just as individual protons and neutrons have a spin of ½, so the nucleus will emhibit a nuclear spin, designated by I, which will reflect the net effect of the nucleons present. It may take fractional values ($^1/_2$, $^3/_2$, etc.) where there are an odd number of nucleons or it may take integral values (0, 1, 2, etc.) where there are an even number of nucleons. All nuclei with an even number of protons and neutrons have zero net spin ($I = 0$), no magnetic moment, and no resonant absorption frequency. All nuclei with an odd number of protons and/or neutrons possess a magnetic moment (μ), which is related to the nuclear spin quantum number (I) by :

$$\mu = \frac{\nu I h}{H}$$

where h is Planck's constant,
ν is the resonant absorption frequency,
H is the magnetic field strength.

The discovery of the method for the measurement of the nuclear magnetic resonance, outlined above, was made in 1945 by two different groups. The Purcell method (Fig. 2.1a) uses a strong uniform magnetic field to obtain partial alignment of the nuclei and a single coil around the sample. The characteristic frequency at which energy is absorbed from the coil is determined. In the other method, known as the Block method (Fig. 2.1b), a strong uniform magnetic field is again used for orientation, but two coils with their axes at right angles are used. Energy absorbed by the nuclei from one coil attached to the oscillator, is re-emitted and induces a small e.m.f. in the second coil which is recorded. In order to detect the resonant absorption position, it is more expedient to vary the strength of the magnetic field slightly with a small supplementary low frequency winding, than to vary the radiofrequency by a small amount. The supplementary field is coupled to the X plates of a cathode tube, and the RF output to the Y plates, in order to display the spectrum. The Purcell method provides a measurement only of the magnitude of the nuclear magnetic moment, whereas the Block method—although electronically more complex—provides in addition, information concerning the algebraic sign of the nuclear magnetic moment, from a study of the phase relationship between the transmitting and receiving coils. A positive moment is inferred as resulting from a resultant net "spinning" positive charge.

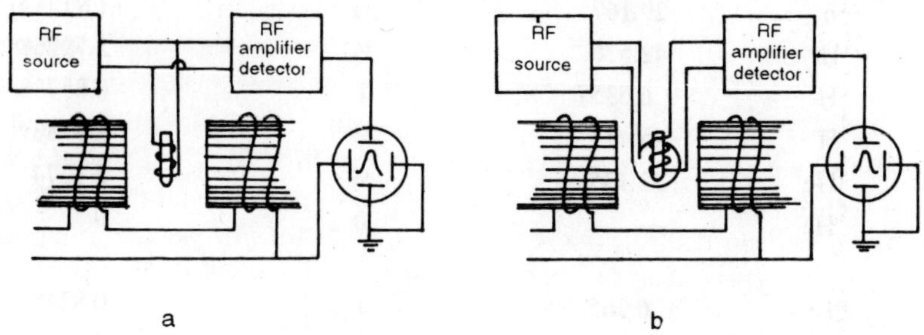

Fig. 2.1 : Measurement of nuclear magnetic resonance.
(a) Purcell Method. (b) Block Method.

Some known nuclear magnetic moments are given in Table 2.1. The value for the neutron which cannot be determined by the above method, has been determined by a modification of an older and more difficult atomic beam method.

The first feature which is apparent is the existence of negative moments—especially that associated with the uncharged neutron, which should have no moment. This anomaly is attributed to the presence of a negative π meson cloud around the neutron. The large positive value for the proton which in theory should be unity, is similarly attributed to the presence of a positive π meson cloud. It will also be noted that the magnetic moment of the deuteron is approximately the algebraic sum of the proton and the neutron, that nuclei with even numbers of neutrons having paired spins and an odd number of protons generally have large moments whilst nuclei with an even number of protons and an odd number of neutrons have small, or negative moments. Particularly striking in this regard, are the relative values of hydrogen-3 and helium-3. Thus a study of nuclear magnetic moments can provide a great deal of evidence concerning nuclear structure. One particular point of importance is the spin value of unity for deuterium. Had this nucleus been a composite of two protons and an electron—an odd number of particles—as seemed necessary in the pre-neutron era, then its spin would have been either ½ or $^3/_2$.

Table 2.1

Isotope	*N.m.r. frequency (Mc/s) for a 10 kilogauss field*	*Spin (I)*	*Magnetic moment (μ) in nuclear magnetons*
${}^{1}_{0}n$	29.167	½	–1.91315
${}^{1}_{1}H$	42.576	½	2.79268
${}^{2}_{1}H$	6.5357	1	0.85738
${}^{3}_{1}H$	45.414	½	2.9788
${}^{3}_{2}He$	32.435	½	–2.1274
${}^{4}_{2}He$	—	0	0
${}^{6}_{3}Li$	6.265	1	0.82192
${}^{7}_{3}Li$	16.547	$^3/_2$	3.2560
${}^{9}_{4}Be$	5.983	$^3/_2$	–1.1773
${}^{10}_{5}B$	4.575	3	1.8005
${}^{11}_{5}B$	13.660	$^3/_2$	2.6880
${}^{12}_{6}C$	—	0	0
${}^{13}_{6}C$	10.705	½	0.70220
${}^{14}_{7}C$	3.076	1	0.40358
${}^{15}_{7}N$	4.315	½	–0.28304
${}^{16}_{8}O$	—	0	0
${}^{17}_{8}O$	5.772	$^5/_2$	–1.8930

Chemical Shifts

Of more particular value to the chemist is a by product of nuclear magnetic studies, now very widely used in the determination of chemical bonding and chemical structures. As mentioned earlier, a sample of water in a magnetic field of 10,000 gauss shows a characteristic absorption at 42,576 Mc/s due to the hydrogen. A similar characteristic resonant absorption frequency is measured not only for water, but for all hydrogen compounds. If a liquid sample is placed in an extremely uniform magnetic field—varying only by one part in a million over the sample—it can be shown that the hydrogen resonance absorption is frequently split. Such studies are usually made by holding the resonant frequency constant using a fixed field, and then applying an additional, variable, very small scanning field. In the case of ethyl alcohol, the resonant line is split into three components (Fig. 2.2) with relative intensities in the ration 3 : 2 : 1 corresponding to the absorption by the methyl, methylene and hydroxyl group hydrogen atoms.

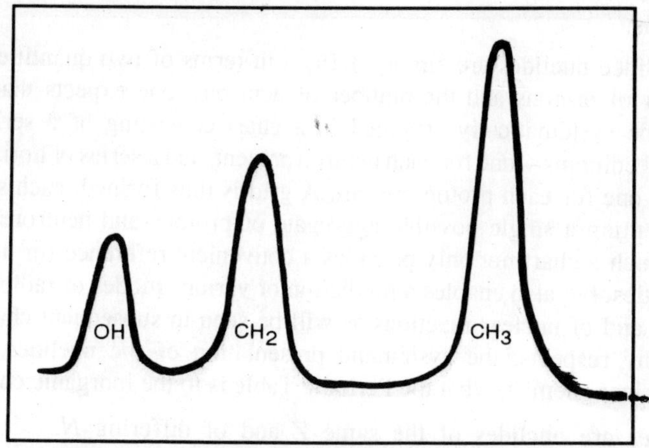

Fig. 2.2 : High resolution nuclear magnetic resonance spectrum for ethyl alcohol.

These slight shifts are due to the fact that the magnetic fields actually experienced by the hydrogen nuclei concerned are not exactly identical to the applied field, which has been slightly altered by the electron clouds associated with the particular groups, *i.e.* with the chemical bond environment.

Commercial high resolution instruments are now widely used in many laboratories to elucidate chemical structures. Ultra high resolution can produce even further splitting, due to the influence of the magnetic moment of neighbouring atoms, but such considerations are beyond the scope of this book, as also is the application of nuclear magnetic resonance to the study of crystal structures and phase transitions to which it has also been applied.

Nuclear Structure

(A) Chart of the Nuclides

The basic building blocks of the nucleus are the nucleons, the protons and neutrons. Each nuclear species, or *nuclide*, can be defined in terms of the number of protons and the number of neutrons within the nucleus. The term isotope is sometimes loosely applied with the same general meaning, but strictly speaking, this term should only be used when discussing two or more nuclides of the same element, *i.e.* nuclides with the same number of protons and differing numbers of neutrons.

Since nuclides are simply defined in terms of two quantities, the number of protons and the number of neutrons, one expects that they could be systematically arranged in a chart consisting of a series of vertical columns—one for each neutron present, and a series of horizontal rows—one for each proton present. A grid is thus formed, each square representing a single possible aggregate of protons and neutrons (Fig. 2.3). Such a chart not only provides a convenient reference for nuclear properties, but also enables a prediction of various modes of radioactive decay, and of nuclear reactions as will be seen in subsequent chapters. In many respects, the systematic presentation of the nuclides, is to the nuclear chemist, what the Periodic Table is to the inorganic chemist.

Isotopes are nuclides of the same Z and of differing N.
Isotones are nuclides of the same N and of differing Z.
Isobars are nuclides of the same mass number A.

It is occasionally found that in a single square—corresponding to a single possible aggregate of protons and neutrons—two or more different nuclear properties are evident, *i.e.* radioactive decay, magnetic moment, etc. This is attributed to a different arrangement of the same nucleons within the nucleus, in different energy states, and the different species are referred to as nuclear *isomers.*

A line of maximum nuclear stability is also normally marked on such a chart—having a slope corresponding to a proton to neutron ratio of 1 : 1 for the lightest nuclides, decreasing to a slope of 1 : 1.6 for the heavy nuclides. For reasons which will be discussed later, the position of a particular nuclide with respect to this line determines the type of radioactive decay to be expected. Nuclides to the right of this line usually decay by β emission while nuclides to the left decay by β^+ emission or by K capture. This distance from the line determines the degree of nuclear instability and consequently the half-life.

(B) Nuclear Stability—Binding Energy

Mutual repulsion between like charged particles, in this case protons, is far greater than the gravitational forces of attraction, and one would therefore expect a nucleus to fly apart spontaneously.

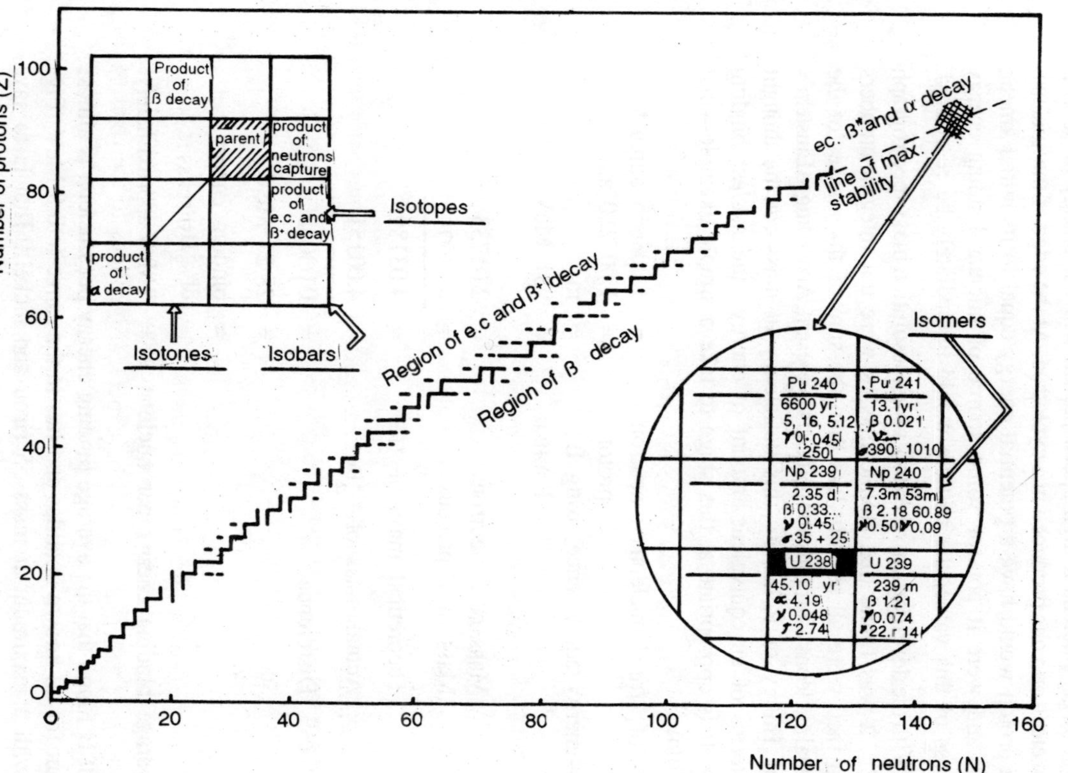

Fig. 2.3 : Plot of the number of protons against the number of neutrons to form a Chart of the Nuclides. Known stable nuclei are indicated by black squares.

It is possible to penetrate the potential barrier of a nucleus, and add to a target nucleus a high speed proton with an energy of several million electron volts produced in a cyclotron. At the edge of the nucleus, such a proton would have a potential energy equal to its initial kinetic energy, and were it not for some source of nuclear binding energy in excess of this value, the particle would immediately be re-ejected.

With the development of Aston's high resolution mass spectrograph in 1925, it was found that atomic masses were not integral numbers and, in fact, were all slightly less than the sum of the masses of the individual nucleons—protons and neutrons—present. According to Einstein's Special Theory of Relativity this deficiency in mass can be thought of in terms of an equivalent amount of energy—the *nuclear binding energy*. It is appropriate at this stage to take a simple example—that of helium.

Masses of "free" nucleons : neutron = 1.008665 a.m.u.*

proton = 1.007280 a.m.u.

Mass—energy equivalence using E = mc2

1 a.m.u. = 931 MeV

Now :

Mass of 2 neutrons = 2.01733

Mass of 2 protons = 2.01456

Theoretical mass of $^{4}_{2}He^{2+}$ = 4.03189

Actual mass of $^{4}_{2}He^{2+}$ = 4.00151

Difference = 0.03038 a.m.u.

= 28.27 MeV

= binding energy of $^{4}_{2}He$

i.e. average binding energy per nucleon = 7 MeV approximately.

It is found that the average binding energy *per nucleon* is almost constant for all known nuclides and lies in the range 6 to 9 MeV (Fig. 2.4), with a maximum at mass 55 (iron and nickel). That is to say, they are the most stable of all nuclei, and it is therefore not surprising that they are the elements most common in meteorites and the earth's core.

*All masses quoted are on the currently accepted scale based on the carbon-12 isotope = 12.000000 a.m.u.

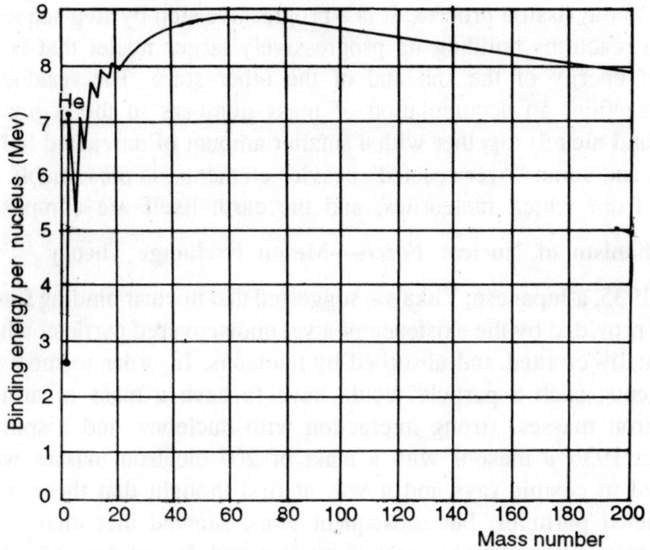

Fig. 2. 4 : Binding energy per nucleon, plotted as a function of mass number.

To understand nuclear binding energy one should think of it as something which the nucleus lacks. For example, if it were possible to bring two protons and two neutrons together to form a helium nucleus, energy equal to the mass loss (28.27 MeV) would be given *out.* The particle would stay together because they were each short of mass equivalent to 7 MeV which they require in order to exist free. Furthermore, they would have to stay together until such time as the required amount of energy was made available to them. The binding energy which is a positive quantity as far as holding the nucleus together is concerned, is in fact, the energy equivalent of the absent mass.

The existence of a maximum binding energy per nucleon in the mid section of the binding energy curve is also the reason why fission of heavy elements and the opposite process, fusion in lighter elements, are both sources of nuclear energy. If we consider, for example, the fission of a uranium-235 or a plutonium-239 nucleus into two roughly equal fragments, then almost 240 nuclear particles have, on an average, increased their binding energies by just under 1 MeV each—or collectively, by some 200 MeV. On the other hand, light nuclei such as deuterium fusing together will also result in an increase in binding energy, which although it is larger on an energy per nucleus basis by comparison with fission, due to the far greater number of nucleons involved in the fission process. It is a progressive step by step sequence of fusion reactions building up progressively larger nuclei that is the source of energy of the sun and of the other stars. The residue of such a reaction, an accumulation of mass numbers in the region of 55 (iron and nickel) together with a smaller amount of unreacted lighter elements and some "over-reacted" heavier elements, is presumably the material from which meteorites, and the earth itself are composed.

(C) Mechanism of Nuclear Forces—Meson Exchange Theory

In 1935, a Japanese, Yukawa, suggested that nuclear binding forces might be provided by the existence of a yet undiscovered particle, which is continually emitted and absorbed by nucleons. In order to meet the requirements, such a particle would have to have a mass of around 200 electron masses, strong interaction with nucleons, and a spin of 0 or 1. In 1936 μ mesons with a mass of 207 electron masses were discovered in cosmic rays and it was at first thought that these were the predicted particles, but subsequent work showed that their other properties were incompatible with those required. It was not until 1947 that the π mesons with a mass of approximately 270 electron masses were discovered and later shown to have the required properties. According to the meson exchange theory, nucleons are envisaged as having a virtual meson cloud which can provide exchange forces at short distances. In this process a meson is exchanged between a pair of nucleons, a π^0 meson for similar nucleons, and charged π^+ or π^- mesons for unlike nucleons which change their identity in the process as is shown diagrammatically in Fig. 2.5.

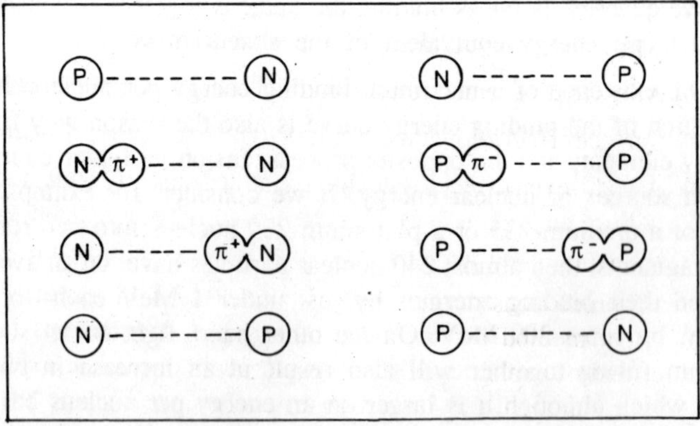

Fig. 2.5 : Exchange of a meson between a pair of nucleons.

(D) Arrangement of Nucleons in the Nucleus

Although there are several theories, the actual structure of the nucleus is not known. Three of these theories will be described briefly here, to give a picture of current ideas to provide a basis upon which to describe nuclear properties.

(I) Liquid drop model

This model which was put forward by N. Bohr in 1936, and further developed by Bohr and Wheeler in 1939, was used in the wartime Manhattan Project. It is largely based on the observation that nuclear volumes are roughly equal to the sum of the volumes of the individual particles. It considers the nucleons as randomly distributed in the nucleus like the molecules in a drop of liquid; any excess energy of one particle is distributing among the others, and spontaneous radioactive decay is rather likened to evaporation of a molecule from the surface after it has acquired sufficient energy from random processes.

With this theory it is possible—with but few exceptions—to calculate the binding energy of a particular nucleus using a semi-empirical equation:

B.E. (MeV)

$$= 14.0\ A - 19.3\frac{(A-2Z)^2}{A} - 0.585\frac{Z^2}{A^{1/2}} - 13.05A^{2/3} \pm \frac{130}{A}$$

This equation takes the form of a summation of five terms each related to a particular nuclear property which effects the binding energy, *viz.*

(*i*) An energy which is proportional to the total number of nucleons.

(*ii*) Allowance for variations in proton to neutron ratio.

(*iii*) Allowance for electrostatic forces of repulsion between protons.

(*iv*) Surface effect allowances for less tightly held surface nucleons.

(*v*) An odd-even effect to account for added stability from nucleon pairing and taken as : + for even proton even neutron nuclei; 0 for odd-even nuclei; — for odd proton odd neutron nuclei.

Correct predictions of the relative stabilities within sets of isobars (constant *A*), a method for the determination of spontaneous fission probabilities and of explaining fission as a process of nuclear excitation followed by elongation and distortion into a dumbell shape prior to capture, are some of the advantages of this theory. A real difficulty is explaining nuclear isomers and abnormally high binding energies associated with nuclei containing certain "*magic*" *numbers* of either protons or neutrons. This led to the development of the nuclear shell model in 1949.

(2) Nuclear Shell Model

An observation that nuclei with certain magic numbers of protons or neutrons *viz.* 2, 8, 20, 28, 50, 82 and 126 show exceptional stability, led to the nuclear shell model which—unlike the liquid drop model—envisages an arrangement of nucleons in energy levels somewhat analogous to the arrangement of electrons in atomic energy shells.

Supporting observations for the existence of magic numbers are:

(*a*) Elements where *Z* is a magic number have a large number of stable isotopes compared with their immediate neighbours.

(*b*) Nuclides where *Z* or *N*—or particularly both—are magic numbers, are most abundant in their mass range.

(*c*) Nuclides with *N* equal to a magic number have very low neutron capture cross-sections, *i.e.* are very reluctant to capture extra neutrons, while nuclides with one neutron less than a magic number have very high cross-sections. Also nuclides with one neutron more than a magic number can—in an excited state—eject a neutron, *e.g.* delayed neutron emission from the fission fragments $^{87}_{36}K$ and $^{137}_{54}Xe$.

(*d*) Three of the four radioactive decay series—all of which end in a different nuclide—terminate in a nuclide where *Z* is a magic number.

(*e*) In α decay to a magic number a high energy α is emitted, whereas α decay from a magic number is of low energy.

In addition to the magic number evidence, there is also the observation that nucleons prefer to be paired in nuclei. For example, of the known stable nuclides there are :

164 nuclides with even *Z* and even *N*
55 nuclides with even *Z* and odd *N*
50 nuclides with odd *Z* and even *N*
4 nuclides with odd *Z* and odd *N*.

The existence of nuclear isomers, nuclides with the same number of both protons and neutrons, yet having different nuclear properties, suggests some sort of internal structural arrangement, possibly in different energy levels, and also lends support to the nuclear shell model.

The above evidence, especially that of the magic numbers associated with particularly stable nuclear configurations (which are reminiscent of the atomic rare gas structure), suggests some system of energy level arrangement, similar to that described for electron orbitals. It is quite apparent that the nuclear magic numbers (2, 8, 20, 28, 50, 82 and 126) are not identical with the total number of electrons associated with the inert gas structures (2, 10, 18, 36, 54 and 86), so therefore any scheme devised to arrive at such numbers must differ accordingly. It is perhaps not surprising that this should be the case, when one considers the physical differences between the two systems. For example, nucleons move in a nuclear force field for which they themselves are responsible and which is characterised by short range forces, instead of the simple central attracting field associated with the electron orbitals. Furthermore, there are both protons and neutrons to be accommodated, each of which presumably occupy their own set of energy levels within the same structure.

Early attempts to manipulate a set of quantum numbers, in a manner analogous to the electronic quantum numbers, were not particularly successful in accounting for observed nuclear properties until 1949. In this year Maria G. Mayer developed a suggestion by Fermi, namely that there may be a tendency with nucleons for there to be a strong association between the orbital angular momentum of the particle and its spin, with a preference for alignment in the same, rather than the reverse, direction.

Briefly, the system which evolved was to employ first a *radial quantum number* (n) which may take values 1, 2, 3, etc. in the same way as the principal quantum number in atomic orbitals, then an *orbital quantum number* ($\imath$) having possible values 0, 1, 2, 3, etc. designated by *s, p, d, f*, etc. by analogy with electronic orbital nomenclature. However, instead of invoking the particle *spin* of ±½ at the end, it is brought forward, and combined with the orbital quantum number, to provide composite *total angular momentum quantum numbers* (j) = $\imath + ½$ and $\imath$ — ½. These replace the single orbital quantum number, ($\imath$) so that, for example, instead of an f subshell ($\imath = 3$) we have two subshells $f_{7/2}$ and $f_{5/2}$. Since there are now two values of j instead of one value of $\imath$, there will be $2j + 1$ states for each value of j corresponding to $2(2\imath + 1)$ sub-divisions of $\imath$ as with atomic orbitals. The system gives rise to the nuclear energy level scheme shown in Table 2.2. This has been drawn in a similar form to that for atomic orbitals.

Table 2.2

Nuclear subshell in order of filling		*Subshell capacity (2j + 1)*	*Shell capacity corresponding to abrupt energy differençe*	*Cumulative total (magic numbers)*
Electronic nomenclature	Mayer nomenclature			
$1s_{1/2}$	$1s_{1/2}$	2	2	2
$2p_{3/2}$	$1p_{3/2}$	4		
$2p_{1/2}$	$1p_{1/2}$	2	6	8
$3d_{5/2}$	$1d_{5/2}$	6		
$3d_{3/2}$	$1d_{3/2}$	4		
$2s_{1/2}$	$2s_{1/2}$	2	12	20
$4f_{7/2}$	$1f_{7/2}$	8	8	28
$4f_{5/2}$	$1f_{5/2}$	6		
$3p_{3/2}$	$2p_{3/2}$	4		
$3p_{1/2}$	$2p_{1/2}$	2		
$5g_{9/2}$	$1g_{9/2}$	10	22	50
$5g_{7/2}$	$1g_{7/2}$	8		
$4d_{5/2}$	$2d_{5/2}$	6		
$4d_{3/2}$	$2d_{3/2}$	4		
$3s_{1/2}$	$3s_{1/2}$	2		
$6h_{11/2}$	$1h_{11/2}$	12	32	82
$6h_{9/2}$	$1h_{9/2}$	10		
$5f_{7/2}$	$2f_{7/2}$	8		
$5f_{5/2}$	$2f_{5/2}$	6		
$4p_{3/2}$	$3p_{3/2}$	4		
$4p_{1/2}$	$3p_{1/2}$	2		
$7i_{13/2}$	$1i_{13/2}$	14	44	126
$7i_{11/2}$	$1i_{11/2}$	12		

There is one rather confusing divergence in the naming of the subshells which arises from the fact that some authors retain the condition that $i \eqslantless n-1$ in complete analogy with electronic orbitals while others—including Maria G. Mayer—do not. Fortunately, this does not effect the outcome of the scheme, other than to provide two possible terms to describe the same thing. Since either of the two systems may to describe the same thing. Since either of the two systems may be encountered, both have been included here, the former under the sub-heading "electronic nomenclature" and the latter under the sub-heading "Mayer nomenclature".

The validity of employing a composite fractional quantum number $j\ (= l \pm \frac{1}{2})$ lies in the agreement which is achieved between theoretical prediction and experimental evidence. The following points support the system :

(*1*) Magic numbers derived from the use of the nuclear shell theory agree with those observed experimentally, and the exceptional stability and high binding energy of nuclei with full shells *e.g.* $^{4}_{2}He$ is explained.

(*2*) Nuclear magnetic moments which are associated with unpaired nucleons, and are zero for even nuclei, should, in the case of nuclei with either an odd proton or an odd neutron, be equal to the *j* value of that unpaired particle. For example, the spin of carbon-13 with an unpaired $p_{1/2}$ neutron should be ½ which it is, whilst oxygen-17, with an unpaired $d_{5/2}$ neutron should be $^5/_2$, which is again the case. In the cast majority of cases the spins of nuclei with an odd nucleon are the same as the *j* value of the unpaired nucleon. In the few cases where this is not so, such as occurs when *j* has a very high value, the spin is invariably that of the next lower *j* value, indicating preferential filling of this alternate subshell. It is also found in such cases, that the nucleus exhibits its theoretical spin when in an excited energy state, *i.e.* in spin measurements of an isomeric state.

(*3*) Nuclear isomers are explained as long-lived excited states of the same nuclide resulting from adjacent closely spaced energy levels which have a large spin difference and would therefore be expected to occur where these spin conditions prevail. This is in fact so. An isomeric transition can be regarded as a measurably slow γ emission involving a large spin change, in much the same way as an electronic triplet (or forbidden) transition involving a spin change leads to the slow phenomenon of phosphorescence compared with the allowed transition which is responsible for the phenomenon of fluorescence in optical spectra. By convention, the isomer corresponding to the upper energy state is indicated by an "m" after the mass number, to indicate that it is the metastable species, *e.g.* the isomers of zinc 69 are $^{69}_{30}Zn$ and $^{69m}_{30}Zn$. Nuclear metastable isomers tend to decay to the ground state with a half-life which may be as short as a fraction of a second or which may run into years. In some instances, where the ground state is itself radioactive, the metastable state may have a half-life considerably longer than the ground state and even undergo a different mode of decay. To take a somewhat extreme example: americium-242 in the ground state undergoes two alternative modes of decay (β^- and electron capture) with a half-life of only 16 hours, while the metastable isomeric state decays by α emission with a half-life of 152 years.

In assessing the value of the nuclear shell model to explain many nuclear phenomena, one must at the same time remember that the theory was in fact tailored to suit known facts. It has not yet led to a quantitative determination of nuclear energy levels nor, for example, does it explain even qualitatively, nuclear asymmetry inferred from nuclear quadrupole measurements. While the nuclear shell model explains some of the things which the liquid drop model cannot, it also has some shortcomings.

(3) Collective model

As a result of the apparent virtues, and shortcomings, of both the above models, A. Bohr—son of N. Bohr—proposed a "collective" model in 1951. This model represented a marriage of the liquid drop and nuclear shell models. Basically, this model adopts the nuclear shell model for the nuclear core, but retains the idea that the surface of the nucleus behaves like the surface of a liquid drop.

The foregoing models are by no means the only ones proposed and probably do not represent the system which will be ultimately adopted. They do, however, provide some picture of nuclear properties and a perspective on current development of nuclear models. Some of the recent experimental findings awaiting explanation are the directional properties of radiations emitted from radioactive nuclei oriented in strong magnetic fields at low temperatures, and the fact that if high energy electron scattering experiments have been interpreted correctly, the charge density of nuclei is not constant, but decreases from the centre out. This applies also to the hydrogen nucleus—the proton. These same experiments also suggest that the helium nucleus is even smaller than the deuterium nucleus.

(E) Representation of Nuclear Energy Levels

Nuclear energy levels are frequently represented as shown in Fig. 2.6a by lines indicating the energy levels in a box or "well" as it is called. The width of the well corresponds to the nuclear diameter and the depth of the well represents the binding energy. The raised edge describes the potential barrier which would be encountered by a charged particle approaching from the outside, and its height is a function of the charge on the bombarding particle. For a neutron with zero charge, it would be non-existent. Such a diagram, which is essentially an energy cross-section of the nucleus, is convenient for describing nuclear bombardment reactions, but is inconvenient for representing simple radioactive decay from one nuclide to another. In this case, a diagram such as that in Fig. 2.6b is preferred. Here, only the ground states and relevant excited levels of the nuclei concerned are represented by lines whose relative vertical displacement indicates their relative energy. By convention, the daughter product of α decay, positron emission and electron capture, *i.e.*, those with an atomic number less than the parent, is shown to the left of the parent, while the daughter product of β^- decay, which has an atomic number one greater than the parent, is shown to the right of the parent.

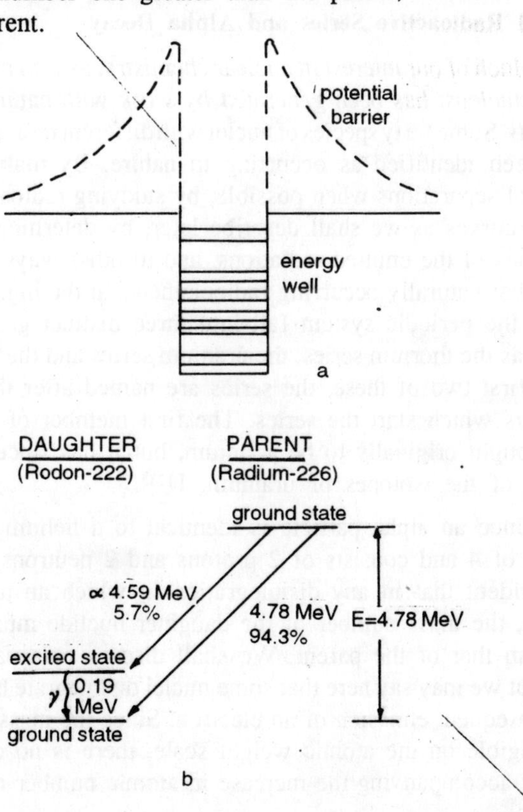

Fig. 2.6 : Methods for the representation of nuclear energy levels.

3
Radioactivity and Nuclear Reactions

Natural Radioactive Series and Alpha Decay

Much of our interest in nuclear chemistry, as well as our knowledge of the nucleus, has been generated by work with naturally radioactive elements. Some forty species of nuclei with different radioactive properties have been identified as occurring in nature. By making physical or chemical separations when possible, by studying radioactive decay and growth curves as we shall describe later, by determining the specific properties of the emitted radiations, and in other ways, scientists have found that naturally occurring radioelements at the high atomic weight end of the periodic system fall into three distinct groups. These are known as the thorium series, the uranium series and the actinium series. In the first two of these, the series are named after the longest-lived members which start the series. The first member of the third series was thought originally to be actinium, but it has since been found to be one of the isotopes of uranium, U^{235}.

Since an alpha particle is identical to a helium nucleus, it has a mass of 4 and consists of 2 protons and 2 neutrons. Consequently, it is evident that in any disintegration in which an alpha particle is emitted, the mass number of the daughter nuclide must be four units less than that of the parent. We shall discuss beta particle emission later, but we may say here that some nuclei disintegrate by the formation and subsequent emission of an electron. Since the mass of the electron is negligible on the atomic weight scale, there is no change in mass number accompanying the increase in atomic number observed in this process.

Standard reference books contain the detailes of the naturally radioactive series, but we will include here, for illustrative purposes, a diagram (Fig. 3.1) showing the decay steps in the uranium series.

There are a few instances of alpha ray decay in elements having a lower atomic number than 83, but all isotopes of elements above this number exhibit either alpha or beta decay, and alpha radioactivity is usually considered to be a characteristic of large nuclei. It mught also be pointed out that many alpha emitters have neutron-proton ratios which would be considered as "stable" from that standpoint, and it is usual to refer to a nucleus as being unstable toward either alpha decay or beta decay.

The results of a hypothetical experiment involving the addition to an alpha particle to a radon nucleus indicate that it would require about 27 MeV to bring a doubly charged alpha particle to a distance of 9×10^{-13} cm from a "point" nucleus of radon. A similar calculation would represent the reverse of this problem. Suppose we were to begin with the Ra^{226} nucleus with its 88 protons and 138 neutrons. Since the protons and neutrons are arranged in levels inside the nucleus, we may think of the radium nucleus as having been formed in the earlier hypothetical experiment by the addition of a helium nucleus to the radon nucleus, and the addition of a completed nuclear level. If this group of two neutrons and two protons were now removed from the radium nucleus, we could determine the amount of energy it should have when it comes out. It was suggested that this alpha particle should come out of the nucleus with the same amount of energy (27 Mev) required to introduce it into the radon nucleus. One of the very surprising results in the early study of radioactive materials was that alpha particles come out of Ra^{226} nuclei with about 5 Mev rather than the 27 Mev expected.

The problem of how an alpha particle is able to come out of the nucleus with less energy than would be required to introduce it cannot be explained except by using a quantum mechanical way of describing the behavior of particles. This is difficult to comprehend in an ordinary sense, but it is the only method which gives an adequate description of the behavior of very small particles. We think of the alpha particle as striking the nuclear barrier a very great number of times, and find that occasionally the alpha particle appears outside the nucleus where it can be detected with suitable instruments. Since a given alpha particle may strike the barrier 10^{30} times before coming out, it is seen that this is a highly improbable occurrence, but methods of detecting these individual nuclear events are extremely sensitive.

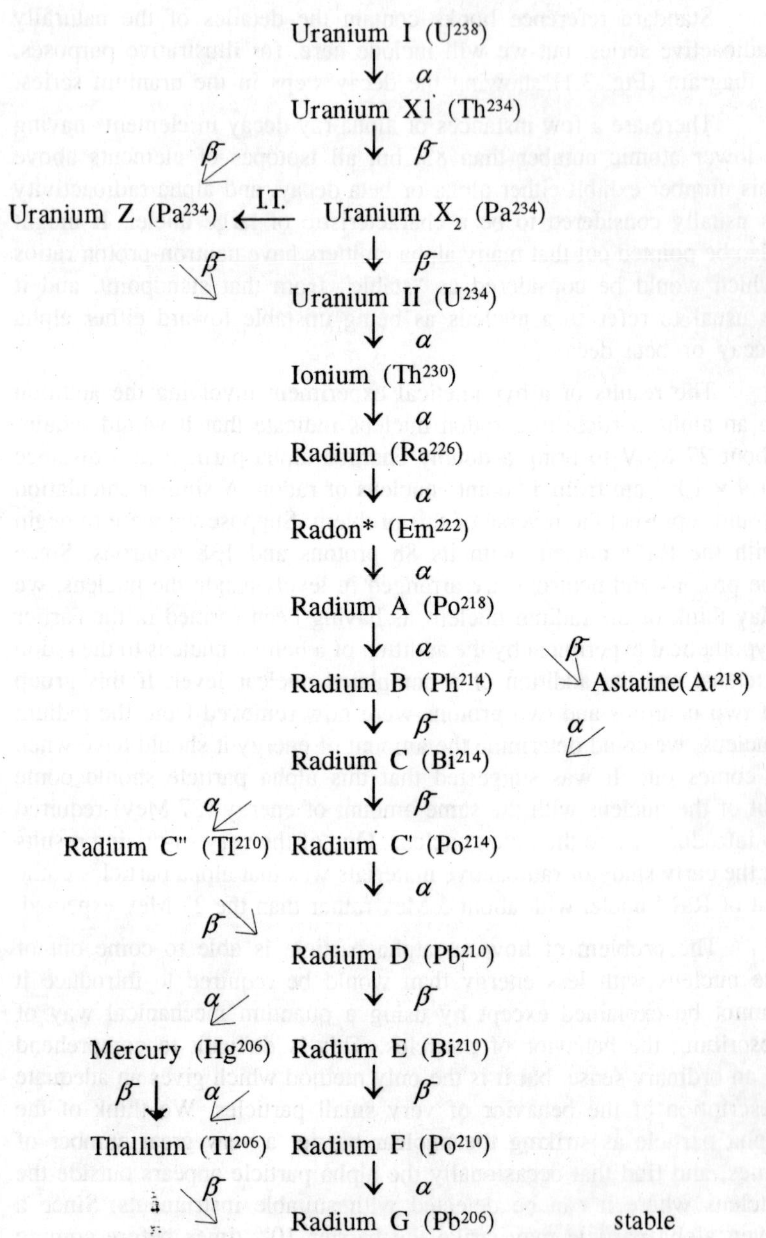

Fig. 3.1 : the Uranium Decay Series.

*Em = emanation, = 86. Radon is the most common isotope of this element.

As difficult as this idea might be to accept, the Gamow "tunneling effect" is our only explanation for alpha particle emission. The question might be asked, "Why do we accept explanations such as this for alpha emission and other experimental observations in the nuclear field?" The answer is two-fold : (1) We have no other explanation based on our classical physical ideas and (2) the mathematical statements of these quantum mechanical concepts permit us to explain, verify, and to predict many new ideas about atomic and nuclear structure and relationships. For example, by using this method of description, we can explain not only the energy of the observed alpha ray emission, but also the probability of its occurrence, which gives rise to the experimentally observable disintegration rate or half life.

Nuclear Bombardment Reactions

A nuclear bombardment reaction is a process where conversion of one nuclide to anoher is achieved by a bombarding particle (or photon). The first transmutation was effected by Rutherford by bombarding nitrogen with α-particles :

$$^{14}_{7}N + ^{4}_{2}He \rightarrow ^{18}_{9}F* \rightarrow ^{17}_{8}O + ^{1}_{1}H$$

It will be noted that the reaction includes an intermediate *compound nucleus*—an isotope of fluorine in an excited state. The compound nucleus theory of nuclear reactions is universal. The target nucleus absorbes the incoming particle to form a compound nucleus, which is invariably in an excited energy level, and which subsequently decays by one or more mechanisms.

Nuclear reactions may be *endoergic* (energy absorbed) or *exoergic* (energy released). The energy of a nuclear reaction is represented by the symbol Q, and if :

(*i*) net Q of a reaction is negative, the reaction is endoergic and energy must be supplied;

(*ii*) net Q of a reaction is positive, the reaction is exoergic and energy will be released.

The value of Q may be determined from a measurement of the energies of the bombarding and emitted particles of photons, or it may be determined from the difference in the masses of reactants and products, using Einstein's $E = mc^2$ relationship.

For example, consider Cockroft and Walton's first artificial transmutation (1932), namely the bombardment of lithium with 0.4 MeV protons to produce pairs of 8.8 MeV α-particles.

$${}^{7}_{3}Li + {}^{1}_{1}H + Q_1 \rightarrow {}^{8}_{4}Be^{*} \rightarrow 2{}^{4}_{2}He + Q_2$$

$Q_1 = 0.4$ MeV

$Q_2 = 2 \times 8.8$—16.7 MeV

Qnet $= Q_2 — Q_1 = +\ 17.2$ MeV (exoergic)

Check of mass balance :

Reactants	= 7.01436 + 1.00728	= 8.02164
Products	= 2 × 4.00151	= 8.00302
Difference		= 0.01862 a.m.u.

1 a.m.u. = 931 MeV ∴ 0.01862 a.m.u. = 17.3 MeV

The reaction energy can also frequently be inferred from other data, for example, an approximate estimate of the vast amount of energy available in fission of, say, ^{239}Pu can be calculated from the binding energy curve. The 239 particles each with an average binding energy of 7.3 MeV, are rearranged so that they each have an average energy of 8.2 MeV. The available energy is therefore (239 × 8.2) — (239 × 7.3) = 200 MeV (approximately).

(A) Cross-Section

The probability that a bombarding particle will produce a nuclear reaction is expressed in terms of *cross-section*. The unit is the *barn* (1 barn = 10^{-24} cm^2). Note that the cross-section is not the geometric cross-section of a nucleus, but is a statement of probability. It is a function, not only of the target nucleus, but also of the type and energy of the bombarding particle. Further, the same target nucleus being bombarded by one type of particle of one energy can be described by more than one cross-section, depending upon the subsequent decay of the compound nucleus. This is best understood by taking a specific example, namely the bombardment of ^{235}U by slow neutrons to form the compound uncleus $^{236}U^{*}$. In about six cases out of seven this undergoes fission, but in the other remaining case, emits its surplus energy as a γ-ray and goes to the ground state of ^{236}U.

We thus specify the neutron cross-section of ^{235}U for the fission reaction as being 580 barns, and the cross-section for the (n, γ) reaction as being 107 barns—the total thermal neutron cross-section being 580 + 107 = 687 barns. Neutron cross-sections vary between wide limits. For example, the low cross-sections associated with the reactor moderator materials deuterium (0.00057 barns) and carbon (0.0037 barns) to the high values of neutron absorbing control rod materials such as cadmium (2500 barns).

For a thin target where there is no significant attenuation of the "flux", the total number of reactions taking place per cm^2 of target per unit time (N) = total number of bombarding particles per cm^2 per unit time (flux Φ) × sum total nuclear cross-section, *i.e.*

$$N = \Phi\sigma\ nx$$

where σ is nuclear corss -section,
n is number of nuclei per cm^3,
x is thickness of target (cm).

The amount of isotope produced is proportional to the duration of the irradiation provided that it is either stable, or has a half-life that is long by comparison with the irradiation time. For short-lived products, a saturation level is reached where the amount being produced in unit time is equal to the amount decaying in unit time, and is therefore, a function of the half-life of the nuclide produced.

(B) Types of Reactions

There are a large number of possible nuclear reactions, depending upon the bombarding particle, and upon the subsequent mode of decay of the compound nucleus formed. The various types of reaction are, by convention, identified by placing the symbol for the bombarding particle (or photon) first, a comma, then the symbol for the emitted particle (or photon). A few of the more important examples are :

(*a*) The neutron induced fission reaction—n, f reaction—in uranium

$$^{235}U\ (n,\ f)\ \text{fission products}$$

(*b*) The production of uranium-239, which, by subsequent β decay leads to plutonium, is an example of a n, γ reaction

$$^{238}U\ (n,\ \gamma)\ ^{239}U$$

(*c*) A reverse of this reaction type is the γ, n reaction, which is the basis of the antimony–beryllium neutron source

$$^{9}Be\ \ (\gamma,\ n)\ ^{8}Be^{*} \rightarrow 2\,^{4}_{2}He$$

(*d*) Other types of neutron source *e.g.* the radium or americium–beryllium neutron sources, are based onthe α, n reaction

$$^{9}Be\ (\alpha,\ n)\ ^{12}C$$

(e) Some Van de Graaff accelerator types of neutron source are based on a d, n reaction

$$^{9}Be\ (d,\ n)\ ^{10}B$$

(*f*) The production of carbon-14, the key to carbon dating techniques, and a valuable tracer in organic reaction studies, is produced by a n, p reaction

$$^{14}N\ (n,\ p)\ ^{14}C$$

Neutron bombardment reactions such as the n, p reaction , to produce carbon-14, are the exception rather than the rule. Usually, neutron irradiation results in a n, γ reaction and leads to an isotope of the elements irradiated but with an atomic mass number one greater than the irradiated nuclide. As the isotope produced has an extra neutron it is very frequently a radioactive β emitting isotope of the target material

The above examples represents only a proportion of nuclear bombardment reactions. although a large number of reactions are possible, those which are most important to the chemist are the n, γ reactions. These are the basis for the commercial production of a very large number of available radiotracers, and also for *neutron activation analysis.*

Neutron activation is an analytical tool of wide application which has rapidly developed over recent years. Because radio-active measuring techniques are sensitive, literally to the point of detecting individual atoms, the method permits accurate assays of very small amounts of elements. The underlying principle is simple, namely to activate a sample by neutron irradiation, and, by measuring the amount of activity produced, to determine the amount of material present. For an element to be suitable for neutron activation analysis, it must meet the following conditions:

(*a*) It must have a sufficiently high neutron capture cross-section to effect a reasonable number of transmutations in a relatively short time with a modest neutron flux.

(*b*) The product of the irradiation should be radioactive, emit a radiation suitable for counting, and have a half-life long enough to count, yet short enough to provide a reasonably high level of activity from the small fraction of active isotope produced. If the activity measurements are made with a γ spectrometer, the simultaneous quantitative estimation of a number of elements in a sample is possible.

Of all the naturally occuring elements, half are suitable for detection and measurement in milligram quantities with a flux of only 10^9 neutrons per cm

In a few selected cases low flux neutron sources of the radium-beryllium or americium-beryllium type can be used but for routine neutron activation analysis a flux in the range 10^8 to 10^{10} n/cm^2, is necessary and neutron sources are generally of the particle accelerator type —either Van de Graaff, or voltage multiplier (Cockroft–Walton). The nuclear reactor may, of course, also be used, although for this application it represents an unnecessarily large capital investment.

A recent and spectacular example of neutron activation analysis was that carried out in 1961. Milligram samples of hair, which had been taken from Napoleon the day after his death (May, 1821) were subjected to neutron activation analysis. The results showed conclusively that Napoleon did not die from natural causes, but was the victim of arsenical poisoning. Not only was it possible to establish the presence and quantity of arsenic, but by scanning the arsenic content of some 13 cm long hairs—representing one year's growth—it was also possible to show that the exposure to arsenic was intermittent during this time, and to determine when the arsenic was administered.

NUCLEAR TRANSFORMATIONS

ORIGIN OF RADIATIONS

The types of nuclear transformation which are discussed here are :

α decay,
β^--decay,
β^+ decay,
electron capture,
γ emission
internal conversion.

There is a certain stable proton-neutron configuration, divergence from which results in nuclear instability and radioactive decay of an unstable *parent* nuclide to a *daughter* nuclide of greater stability.

There are three possible situations that can lead to instability. These are :

(A) Too Many Protons and Neutrons (i.e. too much mass)

In this case there is a tendency to eject mass in the form of an *α-particle*. One might well ask why emit an α-particle comprising from nucleons at one time, rather than omit individual nucleons? The answer lies in the nuclear binding energies. Reference to the binding energy curve shows that the total binding energy corresponding to a mass number of 200 is $7.8 \times 200 = 1560$ MeV and the energy corresponding to mass number 240 is $7.4 \times 240 = 1780$. Thus, in the high mass region, the additional 40 nucleons have a total binding energy of 1780—1560 = 220 MeV, or an average of 5.5 MeV each. Hence to liberate one nucleon would require the supply of 5.5 MeV, and to liberate four separate nucleons would require 22 MeV. The helium nucleus (α-particle) has, itself, a binding energy of 28 MeV, or a surplus energy of 6 MeV. In other words, the liberation of a single nucleon would require the supply of approximately 5.5 MeV, but the liberation of an α-particle would be accompanied by a surplus of approximately 6 MeV, which appears in the form of the kinetic energy of the particle.

(B) Too Many Neutrons for the Number of Protons

In this case a neutron is converted into a proton, a *β-particle*, and a neutrino

$$n \rightarrow p + e^{-}\nearrow + \nu\nearrow$$

In the process, the atomic number is increased by one, and an electron is subsequently absorbed by the outer electron shell to preserve electrical neutrality. The overall atomic mass is unaltered, except for the very small decrease associated with an increase in the nucelar binding energy.

Again, one might ask, why not simply eject a neutron ? There are two ways of answering this. Firstly, the above process not only reduces the number of neutrons by one, but simultaneously increases the number of protons by one. Secondly, there is again the energy consideration. The ejection of a neutron would require the supply of some 5 to 8 MeV of energy (the binding energy of the neutron) while the electron has a rest mass of only 0.51 MeV to be supplied by the nuclear rearrangement. This energy, together with the surplus energy in the form of kinetic energy is in fact, the difference between the total binding energy of the parent and daughter nuclides. Attention should also be drawn to the reason for including a neutrino in the above equation. This is necessary for two reasons. Firstly, β-particles are not all emitted with the same energy. There is a continuous distribution of energies up to a maximum value, which corresponds to the unexpected energy of the transformation. The energy unaccounted for must therefore be associated with another particle. A second reason is that in writing nuclear equations, mass-energy, charge and *spin* must balance on both sides. Without the extra particle the spins could not be balanced.

Since most artificial isotopes are produced in a nuclear reactor by the addition of a neutron to a stable target nuclide, it follows that such isotopes have a neutron excess and are therefore almost invariably β^- active.

(C) Too Many Protons for the Number of Neutrons

Proton emission does not occur for energy reasons similar to those described above and either of two alternative mechanisms is possible.

(*i*) *Positron emission* resulting from the conversion of a proton into a neutron, positron and neutrino :

$$p \rightarrow n + b^{+}\nearrow + \nu\nearrow, \text{ or}$$

(*ii*) *Electron capture* (sometimes referred to as *K* capture) by the nucleus of one of the electrons from one of the inner atomic *K*, or less frequently, *L* shells :

$$p + e^{-} \rightarrow n + \dot{\nu}\nearrow$$

Either one of the two above modes of decay leads to the same daughter product with an atomic number of one less, but same mass number as the parent. In the former mechanism however, a positron is emitted and an electron is subsequently lost from the outer shell, whereas in the latter case no mass is emitted from the atom. The former mechanism can consequently only proceed if the energy difference between the parent and daughter exceeds 1.02 MeV, the rest mass of 1 electron + 1 positron. If the energy available is less than 1.02 MeV only electron capture occurs, but if it is greater than 1.02 MeV, then both mechanisms are possible.

Electron capture can be observed, even though no detectable particle is emitted, by the X-ray produced when the *K*-shell (or *L*-shell) hole is filled by an outer electron.

The process of electron capture is of special interest, since it is the only mode of decay wherein the rate of decay has been influenced by chemical composition. The half-life of ^{7}Be in BeF_2 is, in fact, 0.08% greater than that in beryllium metal.

γ-ray emission

In the above radiactive decay mechanisms the daughter nuclide is frequently left in an excited energy level. It almost immediately gives up this excitation energy to reach the ground state by the emission of one or more γ-rays, which have energies corresponding to the energy levels of the *daughter* produced in the preceding decay mechanism.

For example in the decay of radium-226, α-particles of 4.78 and 4.59 MeV are emitted. The larger energy is the transition energy between the parent and the daughter ground states, while the lower energy represents decay to an excited daughter level and should be followed by a γ-ray of energy : 4.78 MeV — 4.59 MeV = 0.19 MeV. The measured γ-energy is 0.187 MeV.

In some instances, where de-excitation is association with a large spin difference, the transition is said to be forbidden, γ emission is delayed, and the nuclide remains in a metastable isomeric form.

As a general rule g emission follows most examples of radioactive decay and it is the exception rather than the rule to find a pure a or b emitter as a result of 100% direct decay to the daughter ground state.

Internal Conversion

The emission of electrons with discrete finite energies—unlike β-particles—has been observed associated with γ emission. In all instances, they have an energy exactly equal to the energy of the de-excitation γ minus the binding energy of the *K* or *L* shell of the daughter produced by the decay. These were originally attributed to an internal photo-electric effect between the de-excitation γ-ray and an electron in the *K* or *L* shell. Present belief, however, is that they result from direct interaction between the nucleus and inner electron shell, and that the de-excitation energy is transferred directly to the orbital electron, resulting in its expulsion. Internal conversion electrons are more common when the de-excitation energy is of the same order as the shell binding energy. The *conversion coefficient* is defined as the ratio of the number of conversion electrons which are ejected, to the number of γ-rays emitted.

As a result of the emission of an extra nuclear electron by internal conversion, a vacancy is left in a *K* or *L* shell which is immediately filled from an outer shell, with the emission of the characteristic X-ray. In some instances this X-ray is replaced by an electron from an *outer* shell in a secondary process. These electrons are called "Auger" electrons and should not be confused with the internal conversion electrons associated with the primary event.

NATURAL OCCURRENCE OF RADIOACTIVITY

A number of radioactive nuclides are found in nature. With the exception of a few formed by the action of cosmic radiation in the atmosphere, notably carbon-14, they are either :

(*1*) Nuclides with half-lives comparable with the age of the earth (~5.10^9 years). Among the more common in this group are:

$${}^{40}_{19}K,\ {}^{87}_{37}Rb,\ {}^{232}_{90}Th,\ {}^{235}_{92}U \text{ and } {}^{238}_{92}U,$$

or

(*2*) They are the daughter products of three radioactive decay chains which begin with a long lived parent (Table 3.1).

Table 3.1
Natural Radioactive Decay Chains

Parent	*Half-life (years)*	*Stable end Product*
Thorium-232	$1.39 \cdot 10^{10}$	Lead-208
Uranium-238	$4.5 \cdot 10^{9}$	Lead-206
uranium-235	$7.07 \cdot 10^{8}$	Lead-207

An example of one of these decay chains, the uranium series, is given in Fig. 3.2.

Much of the early work, particularly on the properties of radiations, was carried out with members of the naturally occurring radioactive chains, especially radium and radon. At the present time the greatest value of these naturally occurring radioactive nuclides is in the field of geochronology.

Quantitative Aspects of Radioactive Decay

Radioactive decay is a purely *random* effect, each atom having a statistical probability that it will decay in a certain time independent of its past history and present environment. The *rate* of decay of a particular nuclide is a decreasing function *only* of the *amount* present, *i.e.*, if there is twice as much present, there will be twice as many disintegrations per unit time. The units used to describe the radioactive decay are the curie (4.7×10^{10} disintegrations are second) and its subdivisions, the milli- and microcurie.

At a given instant consider N atoms present, dN of which decay in the time interval dt.

Then

$$\frac{-dN}{dt} \propto N$$

N.B. The rate of decay has a negative value since it is a decreasing quantity.

or

$$\frac{-dN}{dt} = \lambda N$$

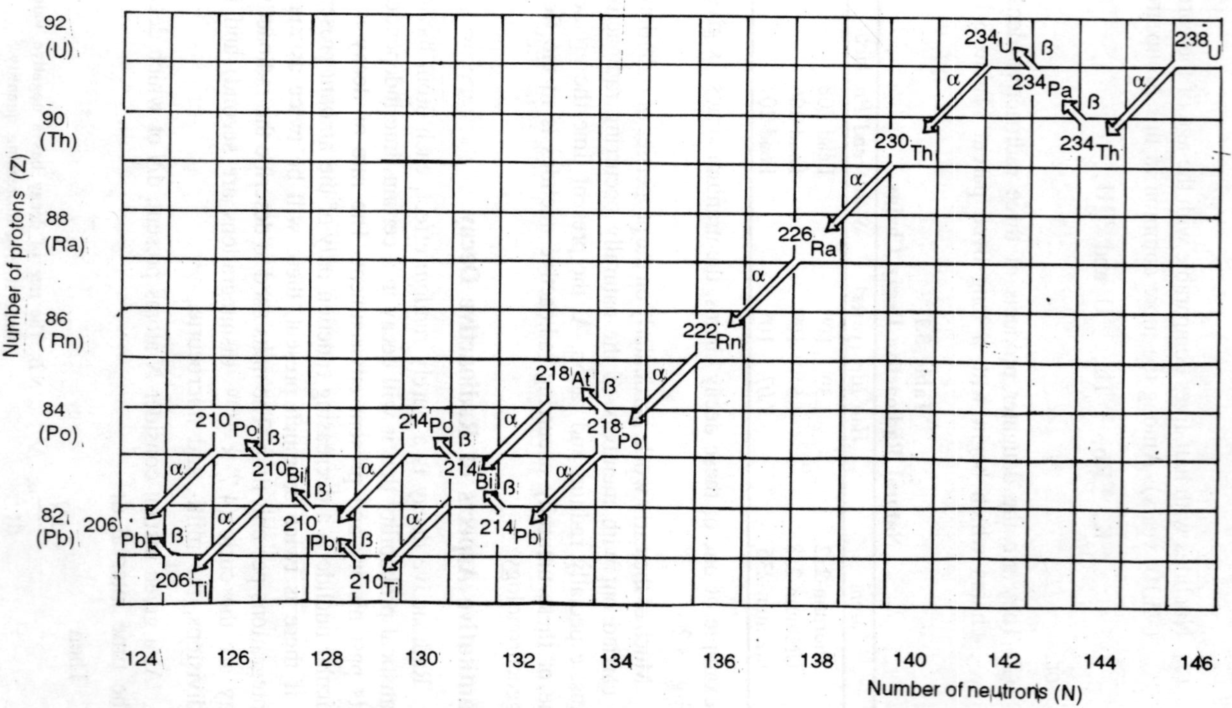

Fig. 3.2 : The Uranium-238 Radioactive Decay Series.

where λ is the *radioactive decay constant* (or disintegration constant) and represents the fraction of those atoms present which decay in unit time

$$\frac{dN}{N} = -\lambda\, dt$$

Integration from time = 0, when N = N_0, to time t, when $N = N_t$ gives:

$$\log \frac{N_t}{N_0} = -\lambda t$$

or

$$Nt = N_0\, e^{-\lambda t} \qquad (3.1)$$

or

$$\log_{10} \frac{N_t}{N_0} = -0.4343\, \lambda t \qquad (3.2)$$

It will be observed that radioactive decay follows an exponential law, and that if the amount of a radioactive nuclide remaining is plotted against time on linear graph paper, an exponential curve similar to that shown in Fig. 3.3a is obtained. From a practical point of view, it is usually more expedient to employ semi-logarithmic graph paper, which gives a straight line plot (Fig. 3.3b). This is more readily extrapolated, and the *half-life* of all but the very short- and the very long-lived nuclides can be directly determined by this means.

Half-life ($t_{1/2}$) *is the time required for half of the original number of atoms present to decay.* Although it is customary to describe a radioactive isotope of an element by its half-life, it is usual to employ the decay constant (λ) in calculations. The decay constant and half-life can be simply related in the following manner :

$$\log \frac{0.5}{1} = -0.4343\, \lambda t_{1/2}$$

i.e.

$$\log 2 = +\, 0.4343\, \lambda t_{1/2}$$

or

$$\lambda = \frac{0.693}{t_{1/2}} \qquad (3.3)$$

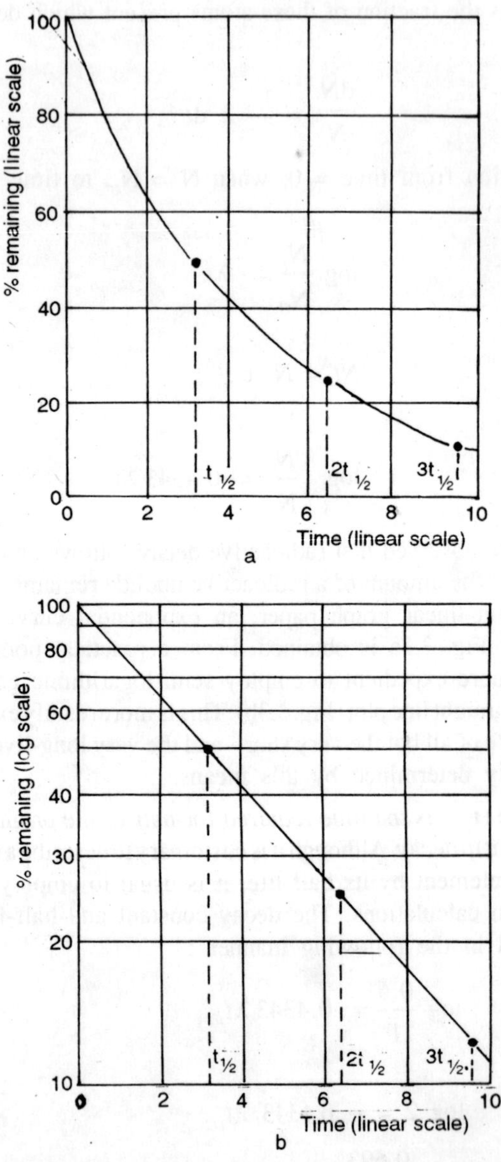

Fig. 3.3 Graphical representation of radioactive decay.

(a) Plotted on linear graph paper.

(b) Plotted on semi-logarithmic graph paper.

Since the amount of radioactive nuclide remaining after 1 half-life is ½ of the original, and after 2 half-lives is ½ of ½ i.e. ¼, it is apparent that the amount remaining after n half-lives is $(½)^n$.

Another parameter frequently employed is *specific activity*—the disintegration rate per unit mass

Specific activity = distintegrations per unit time per unit mass

$= \lambda N$ where N is number of atoms in unit mass

$$= \frac{0.693}{t_{1/2}} \cdot \frac{6.03 \cdot 10^{23}}{A}$$ dis/sec/g (where A is the mass number).

The term specific activity is often widely employed to cover, not just one particular isotope of an element, but a mixture of isotopes. For example, one speaks of low and high specific activity cobalt, to imply either a large or a small dilution of active cobalt-60 with inactive cobalt-59.

(A) Solving Problems

Problems involving radioactive decay can be solved by one of two methods :

(*i*) *Arithmetically.* By converting the information, if given in terms of half-life, into decay constant using eqn. (3.3), and then substituting in eq. (3.2) [or (3.1)]. This method is obviously the most direct, for simple straightforward problems.

(*ii*) *Graphically*, Using semi-logarithmic graph paper. While this method is limited in its numberical accuracy, it is frequently more direct and descriptive than the arithmentical appraoch. As an example, let us estimate the time which has elapsed since the formation of the earth. Thermonuclear processes which go on in the starts to produce cosmic material, produce approximately equal amounts of nuclei of the same mass number. We know from experiment that the half-life of ^{235}U is $7.07 \cdot 10^8$ years, the half-life of ^{238}U is $4.5 \cdot 10^9$ years, and the present ratio of ^{235}U to ^{238}U is 7 to 1000. The problem is to determine how long it is since the faster decaying ^{235}U and the slower decaying ^{238}U were present in equal amounts. To solve the problem graphically, it is only necessary to place points on the graph to indicate the present relative isotopic ratio, to draw from these points, lines having a slope corresponding to the respective half-lives, and to project these lines until they itnersect (Fig. 3.4). The approximate age of the earth thus determined is 5.8×10^9 years. It is also of interest to note that uranium isotope ratios are the same in meteorites as on the earth, indicating a similar age and therefore a similar origin.

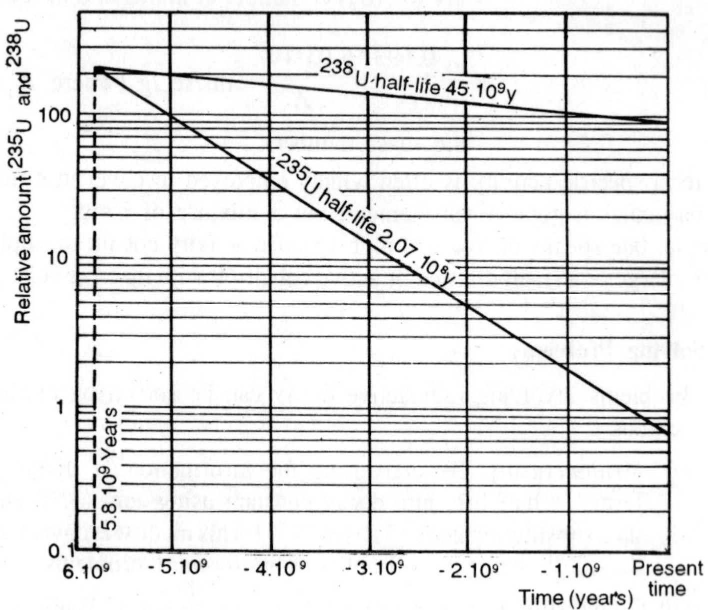

Fig. 3.4 : Determination of the age of the earth by graphical means.

(B) Growth of Radioactive Daughters. Radioactive Equilibrium

Frequently, radioactive decay leads to the production of a radioactive daughter. Of particular interest is the situation leading to *secular equilibrium* where the half-life of the parent is very much greater than that of the daughter.

Consider the case where X decays to Y which in turn decays to Z. Let λ_X and λ_Y be the respective decay constants.

$$X \xrightarrow{\lambda_X} Y \xrightarrow{\lambda_Y} Z$$

e.g. $^{238}U \rightarrow {}^{234}Th \rightarrow {}^{234}Pa \rightarrow$ etc.

Equilibrium is established between X and Y when the number of Y atoms decaying per unit time is equal to the number of Y atoms formed per unit time, which is of course, the number of X atoms decaying per unit time.

Number of atoms of Y decaying $= N_Y\lambda_Y$
Number of atoms of Y forming $= N_X\lambda_X$
as equilibrium $N_Y\lambda_Y = N_X\lambda_X$

i.e.

$$\frac{N_Y}{N_X} = \frac{\lambda_X}{\lambda_Y}$$

or remembering that

$$\lambda = \frac{0.693}{t_{1/2}}$$

$$\frac{N_Y}{N_X} = \frac{t_{1/2 Y}}{t_{1/2 X}}$$

Note

(i) At equilibrium, the ratio of the ratio of the numbers of atoms present is the same as the ratio of the half-lives.

(ii) In the case of all modes of decay where the mass number is constant, the ratio of the masses present is also the ratio of the half-lives, but where there is a change in mass number (*A*), *e.g.* α decay, allowance must be made :

$$\frac{\text{Mass X}}{\text{Mass Y}} = \frac{t_{1/2 X} \cdot A_X}{t_{1/2 Y} \cdot A_Y}$$

(iii) Extension to successive decay, as for example, in the uranium natural radioactive decay series :

$$^{238}_{92}\text{U} \xrightarrow{\alpha} {}^{234}_{90}\text{Th} \xrightarrow{\beta} {}^{234}_{91}\text{Pa} \xrightarrow{\beta} {}^{234}_{92}\text{U} \xrightarrow{\alpha} {}^{230}_{90}\text{Th} \xrightarrow{\alpha} {}^{226}_{88}\text{Ra} \xrightarrow{\alpha} \text{etc.}$$

Although many of the half-lives are long compared with their immediate precursor, they are all very short by comparison with the first member ^{238}U, and therefore in a piece of uranium ore, in which equilibrium has had time to be established, *all* members of the series are in radioactive equilibrium with the parent ^{238}U, and therefore with one another. That is to say, there is the same amount of *activity* from ^{230}Th as there is from ^{234}Pa, or from the parent ^{238}U. The individual *masses* of the members will, of course, differ depending upon their half-life and mass number. Consider, for example, the amount of ^{230}Th in equilibrium with ^{238}U in uranium ores

$$\frac{\text{Mass}\ ^{230}\text{Th}}{\text{Mass}\ ^{238}\text{U}} = \frac{8.0\times10^{4}}{4.5\times10^{9}}\cdot\frac{230}{238} = 1.7\times10^{-5}$$

i.e. there will be 17 mg of ^{230}Th per kg of uranium in uranium ores.

Approach to Equilibrium

The rate at which a long-lived parent establishes equilibrium with a short-lived daughter is, somewhat surprisingly, a function of the half-life of the daughter, and not of the parent. The quantity of a daughter produced as a function of time from a radioactive parent when the daughter produced is itself decaying, involves solving simultaneous differential equations. It is both easier and more instructive to consider the situation from first principles :

Consider a case of two bottles, A and B, containing identical amounts of uranium which have had sufficient time to establish equilibrium with the first daughter ^{234}Th.

$$^{234}_{90}\text{U} \xrightarrow[t_{1/2}\,=\,4.5\cdot10^{9}\,\text{years}]{\alpha} {}^{234}_{90}\text{Th} \xrightarrow[1/t\,=\,24.1\,\text{days}]{\alpha}$$

Treat each bottle in the following way :

(*i*) Leave bottle A alone.

(*ii*) Chemically separate the uranium and thorium in bottle B, return the uranium fraction to bottle B, and place the thorium in a separate bottle b (Fig. 3.5).

The untouched bottle A, is in equilibrium and will remain so. The total combined amount of uranium and thorium in the other pair of bottles will be the same as that in bottle A, since chemical processes and environment do not alter the ultimate course of radioactive decay.

Consider the situation in 24.1 days time. Only half of the equilibrium amount of separated thorium in b is left. Therefore another half must have grown in B. Similarly, after 48.2 days, only $^1/_4$ of the original thorium is to be found in b and therefore $^3/_4$ must be in B; and after 72.3 days, $^1/_8$ will be in b and $^7/_8$ in B. Thus we can construct a thorium growth curve, which is in fact, an inverted decay curve (Fig. 3.6).

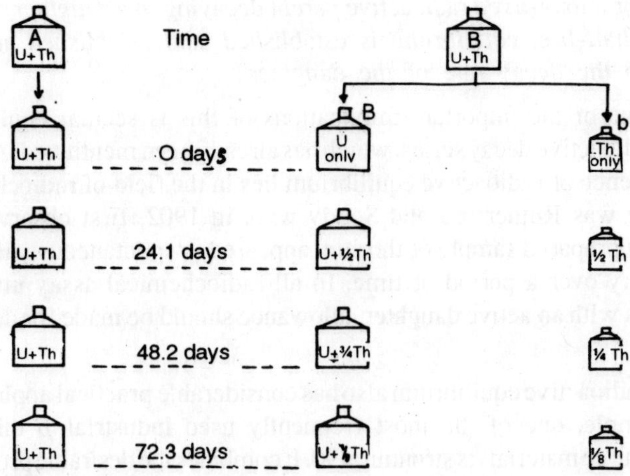

Fig. 3.5 : Diagrammatic representation of the establishment of radioactive equilibrium between a long-lived parent and a short-lived daughter.

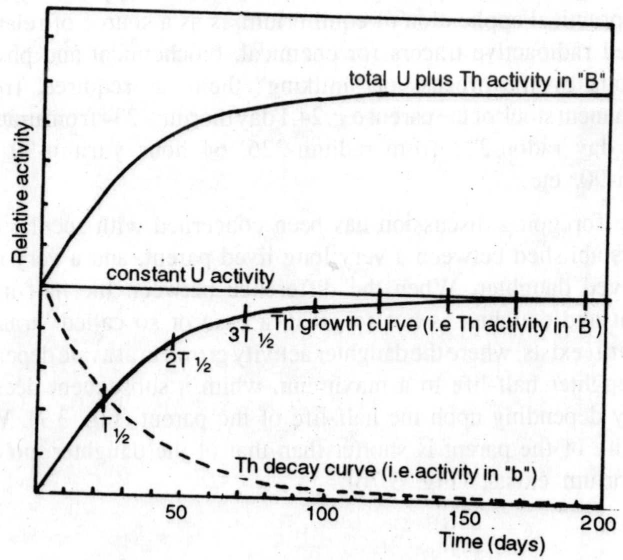

Fig. 3.6 : Radioactive equilibrium growth curve.

For a long-lived radioactive parent decaying to a daughter of much shorter half-life, equilibrium is established, and established at a rate equal to the decay rate of the daughter.

One of the important implications of this is secular equilibrium in the radioactive decay series, which has already been mentioned. Another consequence of radioactive equilibrium lies in the field of radiochemical assay. It was Rutherford and Soddy who, in 1902, first observed that a freshly prepared sample of thorium appeared to spontaneously increase in activity over a period of time. In all radiochemical assay involving a species with an active daughter, allowance should be made for daughter activity.

Radioactive equilibrium also has considerable practical application; for example, one of the most frequently used industrial β thickness gauge source materials is strontium-90. It combines the desirable properties of low γ background with the two apparently contradictory properties of long half-life and high β energy. In fact, the high 2.2 MeV β-particles arise, not from the strontium-90, which emits βs of only 0.61 MeV, but from the yttrium-90 daughter, which is in equilibrium with the strontium-90. In other words, the short-lived yttrium provides the βs and the long-lived strontium provides a constant supply of yttrium. Another practical application of equilibrium, is as a source of relatively short-lived radioactive tracers for chemical, biochemical and physical applications by the process of "milking" them, as required, from a semipermanent stock of the parent *e.g.* 24.1 day thorium-234 from uranium-238, 3.8 day radon-222 from radium-226, 64 hour yttrium-90 from strontium-90, etc.

The foregoing discussion has been concerned with secular equilibrium established between a very long-lived parent, and a very much shorter-lived daughter. When the difference between the half-life of the parent and daughter is not so great, a case of so called "*transient equilibrium*" exists, where the daughter activity grows in at a rate dependent on the daughter half-life to a maximum, whith a subsequent decrease in activity depending upon the half-life of the parent (Fig. 3.7). When the half-life of the parent is shorter than that of the daughter, no form of equilibrium exists, (Fig. 3.7b).

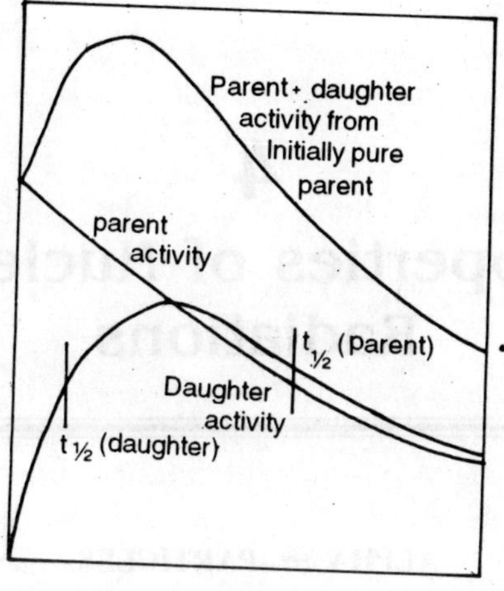

a

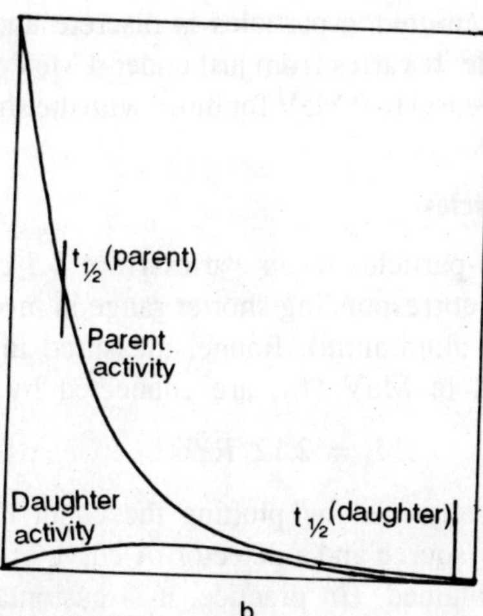

b

Fig. 3.7 : (a) Example of transient equilibrium.
(b) Example of where there is no equilibrium

4

Properties of Nuclear Radiations

ALPHA (α)-PARTICLES

(A) Energy Spectrum of α-particles

The energy of emitted α-particles is discrete and characteristic of the decaying nuclide. It varies from just under 4 MeV for the longest-lived emitters ($\sim 10^{10}$ years) to 9 MeV for those with the shortest observed half-lives ($\sim 10^{-6}$ sec).

(B) Range of α-particles

The range of α-particles in air varies from 2.5 cm (4 MeV) to 9 cm (9 MeV) with a corresponding shorter range in more dense media (0.02 to 0.06 mm in aluminium). Range, measured in cm in air (R, and energy measured in MeV (E), are connected by the relation :

$$E = 2.12\ R^{2/3}$$

The range can be determined by plotting the count rate against the distance between an α source and a detector. A curve of the type shown in Fig. 4.1 will be obtained. (In practice, it is customary to vary the air pressure over a fixed distance to avoid inverse square law effects). Such a curve is observed to have a tailing or straggling at the end. This is not due to small variations in the initial α-particle energy, but to two other factors.

(a) The statistical nature of the collisions which result in the loss of energy by the α-particles.

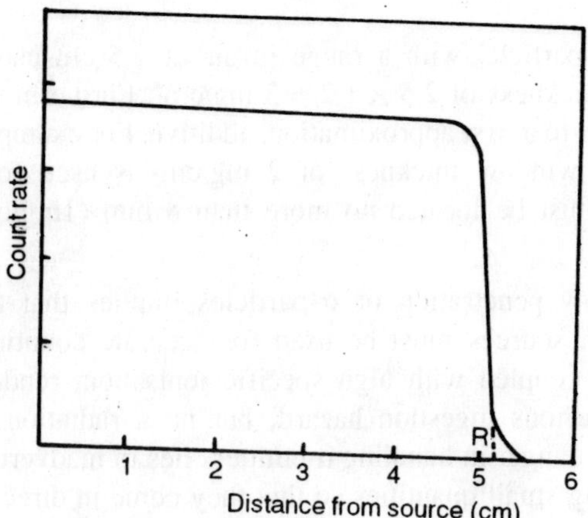

Fig. 4.1 : Range of an α-particle in air.

(b) The attachment of electrons, one at a time, so that toward the end of its path, an α-particle exists as He+, before the final electrón is attached.

The observation that range, and hence energy, are related to half-life, led Geiger and Nuttall in 1911 to look for a relationship amongst the naturally occurring α emitters. They were able to show that if the logarithm of the range was plotted against the logarithm of the radioactive decay constant, a series of straight lines is obtained which can be expressed by :

$$\log \lambda = \text{A} \log R + \text{B}$$

where λ is radioactive decay constant,
R is α range in air,
A is a constant defining slope,
B is a constant defining the line.

Equivalent thickness. The range of an α-particle (or any other nuclear charged particle) is a function of the density of the material in which the range is measured, and therefore direct comparison of range in different media is difficult. The situation is simplified, however, by measurement of ranges in terms of *equivalent thickness*, which has units of mg/cm² and is in fact, an expression of the amount of mass traversed.

Equivalent thickness (mg/cm²) = actual thickness × density × 1000. Since the density of air at N.T.P. is 0.0012 g/cc (*i.e.* 1.2 mg/cc), 1 cm of air has an equivalent thickness of 1.2 mg/cm² and therefore a 4 MeV α-particle, with a range in air of 2.5 cm can penetrate an equivalent thickness of 2.5 × 1.2 = 3 mg/cm². Furthermore, equivalent thickness are, to a first approximation, additive. For example, if a Geiger tube with a window thickness of 2 mg/cm² is used for a counting, the source must be located no more than 8 mm (1mg/cm²) from the window.

The low penetration of α-particles implies that thin, virtually "weightless", sources must be used for accurate counting work. The short range, coupled with high specific ionisation, renders α-particle emitters a serious ingestion hazard, but no a radiation hazard. That is to say, the danger in handling α emitters lies in inadvertently inhaling or swallowing small quantities so that they come in direct contact with vital organs.

C) Interaction of α-particles with Matter

Electrostatic interaction with orbital electrons results in ionisation, and the production of *ion pairs*, as the positive ion and ejected electron are called. The average energies expended in producing an ion pair in some common gases are given in Table 4.1. It will be noticed that the energy is about twice the value which is indicated from spectroscopic measurements. This is due to the fact that about half of the energy dissipated goes in producing excited species.

Table 4.1

Gas	*Average energy per ion pair (eV)*
Nitrogen	35.0
Oxygen	32.3
Argon	25.4
Hydrogen	33.0

For air, an average value of 34 eV is expended per ion pair, so the total number of ion pairs produced by a typical 5 MeV α would be :

$$\frac{5 \cdot 10^6}{34} = 1.5 \cdot 10^5 \text{ ion pairs}$$

Not all ion pairs are produced directly by the α-particle. In fact only half of the ionisation which occurs is due directly to an a interaction, the remaining half is secondary ionisation produced by the electrons ejected during the primary α interactions. Furthermore, the degree of ionisation is not constant over the α path. The *specific ionisation,* which is the number of ion pairs produced per mm of path length, is inversely related to the particle's residual energy. This is not as surprising as it may at first seem, because the higher energy αs, which move at a higher velocity, remain in the vicinity of an orbital electron for a shorter time. The specific ionisation varies from approximately 2000 ion pairs per mm at high a energies up to a maximum value of nearly 700 ion pairs per mm at about half a centimeter from the end of their range when their residual energy is only about 1 MeV (Fig. 4.2).

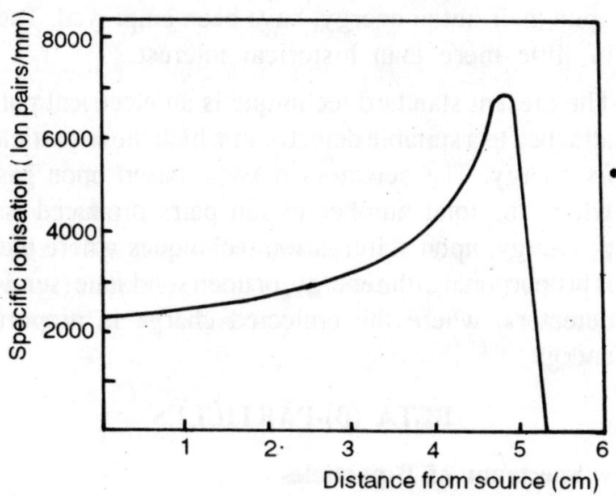

Fig. 4.2 : Specific ionisatioı. of an α-particle as a function of the distance from its source.

Secondary effects, resulting from ionisation, are chemical decomposition and other ionisation induced reactions. These will be dealt with in a subsequent chapter on radiation chemistry. Another secondary effect resulting from electronic excitation is the production of visible radiation (fluorescence) as vacant electron orbitals are refilled. This can be quite spectacular when handling solutions of high specific activity. It is also of considerable application in the detection and measurement of α and other radiation using scintillation techniques.

(D) Measurement of *α* Energy

(a) Since the range and energy are related, an obvious way to determine α energy is to measure the range. A rough determination can be made simply by moving a thin end window Geiger tube away from a source, and at the same time making

allowance for the tube window thickness. A complication associated with this method is the gradual decrease in count rate due to the inverse square law effect. A modification to this technique which can provide reasonably good results with simple equipment, is the use of a detector—preferably a scintillation type—and source at a fixed distance in an air tight container, and to observe the count rate as a function of air pressure.

(b) Magnetic spectrograph methods where the α-particles are deflected by a magnetic field into arcs of varying radius, depending upon their initial energy, have been employed. These are now of little more than historical interest.

(c) The present standard technique is an electrical pulse analyser attached to a suitable detector in which the α-particle dissipates its energy. The detectors may be based upon gas ionisation where the total number of ion pairs produced is a function of energy, upon scintillation techniques where the light flash is proportional to the energy, or upon solid state (semi-conductor) detectors, where the collected charge is proportional to α energy.

BETA (β)-PARTICLES

(A) Energy Spectrum of β-particles

Beta-particles are not emitted with fixed discrete energies as are α-particles. Instead, they exhibit a continuous distribution of energies up to a maximum energy (E_{max}), which is characteristic of the nuclide concerned. Fig. 4.3 shows a typical β energy spectrum. The energy possessed by the largest number of β-particles lies between ¼ and ½ of the value of E_{max}, which is the transition energy of the decay reaction. The difference in the energy exhibited by a particular β-particle and E_{max} is the energy of the neutrino which is simultaneously emitted.

Beta decay energies are considerably lower than a decay energies, and lie between 0.02 and 4 MeV, with the majority in the range 0.5 to 2 MeV. A rule, formulated by Sargent in 1933, relates the decay constant λ to E_{max} in a manner analogous to the Geiger-Nuttall rule for αs. It shows that the plot of log λ against log E_{max} gives points lying on one or another of several straight lines. Each line in fact, depends upon a different "spin" transition between parent and daughter.

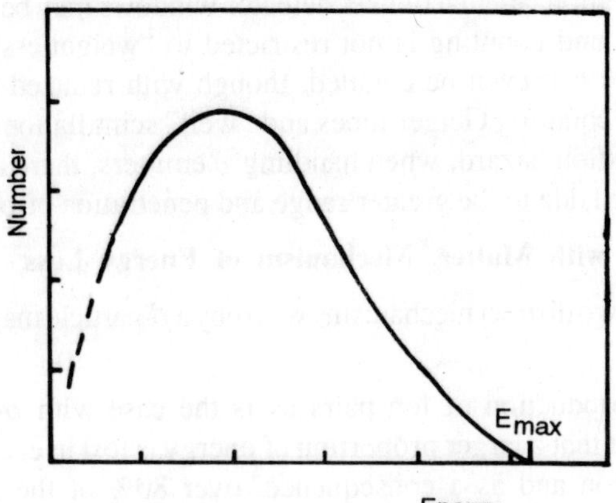

Fig. 4.3 : β-particle Spectrum

(B) Range of β-particles

The range of β-particles is some 500 times that of α-particles of similar energy, varying from 150 cm in air for a β of 0.5 MeV energy, to 850 cm in air for 2 MeV energy. The specific ionisation is correspondingly less, and has a value of the order of 10 ion pairs per mm. The range of β-particles in other material is also correspondingly greater (Fig. 4.4) than that of αs. Some consequences of this longer

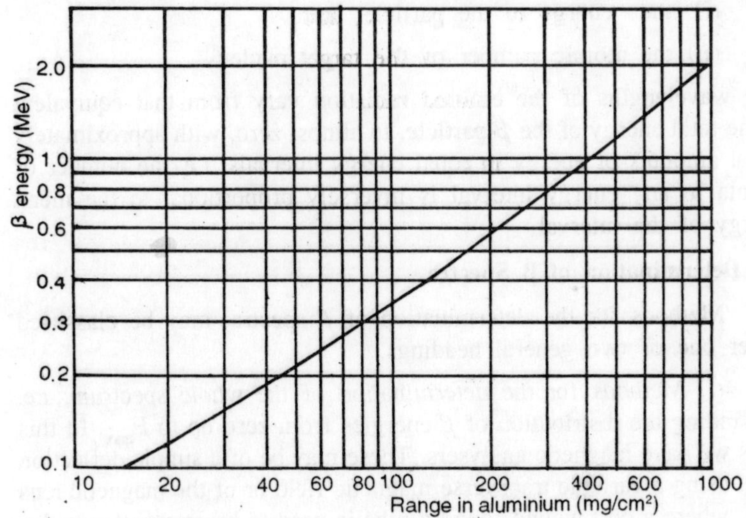

Fig. 4.4 : Range of β-particles in aluminium as a function of their energy.

range and penetration, are that thicker detector windows may be employed with β-particles and counting is not restricted to "weightless" sources. Aqueous sources can even be counted, though with reduced efficiency in special liquid counting Geiger tubes and "well" scintillation detectors. Besides an ingestion hazard, when handling β emitters, there also exists a radiation hazard due to the greater range and penetration of β-particles.

(C) Interaction with Matter. Mechanism of Energy Loss

There are two distinct mechanisms whereby a β-particle may dissipate its energy :

a) The production of ion pairs as is the case with α-particles, except that a larger proportion of energy is lost in each primary collision and as a consequence, over 80% of the ionisation is secondary.

b) Electromagnetic radiation in the X-ray and γ-ray region of the electromagnetic spectrum is produced by the interaction of fast moving β-particles near the positive field of the nucleus, accompanied by a corresponding decrease in the particle's energy—hence the name *bremsstrahlung*, meaning slowing down radiation.

The probability that the energy of β-particles will be converted into bremsstrahlung is proportional to :

(i) the energy of the particle, and

(ii) the atomic number of the target nucleus.

The wavelengths of the emitted radiation vary from that equivalent to the total energy of the β-particle, to almost zero, with approximately equal amounts of energy in equal energy intervals, *i.e.* the number of quanta in any energy interval is inversely proportional to the mean energy of the interval.

(D) Determination of β Spectra

Methods for the determination of β spectra, may be classified under one of two general headings.

(i) Methods for the determination of the whole spectrum; i.e. for finding the distribution of β energies from zero up to E_{max}. In this class we have magnetic analysers. These may be of a simple deflection type, using a variable transverse magnetic field or of the magnetic lens type, where a coil wound round a tube is used to bring the β-particles to a focal point further along the tube. Also in this class of spectrum

determination are electrical pulse analysers coupled to a suitable detector in which the β-particles dissipate all of their energy in the production of ion pairs in a gas, scintillations in a "phosphor", or electron–hole pairs in a solid state detector.

(ii) Methods which lead only to the value of E_{max}. Although limited in that they provide only the value of E_{max}, and not the spectral distribution of energy, these methods are generally of more practical importance since it is the value of E_{max}, rather than the shape of the spectrum, which identifies a β emitter. These methods all involve the use of aluminium "absorbers" to attenuate the count rate, but differ in the method of interpretation. Some of the methods which are employed are :

(a) Graphical representation of count rate as a function of absorber thickness, with extrapolation to zero count—*i.e.* maximum range—and hence the energy from Fig. 4.4. This method is time-consuming and, because of extrapolation errors, is only approximate.

(b) Determination of the thickness necessary to halve the count rate ("half-thickness"). This is quick, but only approximate.

(c) Harley–Hallden method, wherein the relative reduction in count rate for various set values of absorber thickness, is compared with that of a known standard for the same absorber thickness.

(d) Comparison of the absorption curve of an unknown β emitter with that of a known one, as a means of determining the maximum range. This is termed the Feather method.

(e) Bleuler–Zunti method, which like *(c)* and *(d)* above, is a comparative method. The difference lies in the fact that the comparison is not between two sets of experimental results, but between a set of experimental results and a set of standard curves. Absorber thicknesses corresponding to count rate reductions of $(\frac{1}{2})^n$ are referred to published curves. In many respects it is a refinement of the half-thickness method. This method generally gives the most accurate results with the minimum of time.

(E) Scattering of β-particles

The scattering of β-particles, which is very much more pronounced than that for α-particles, is of considerable importance in the radiochemical assay of β emitters. A relatively large proportion of β-particles are, in fact, capable of being reflected back along their original path. This is called *back-scattering*.

The proportion of back-scattered β-particles is a function of the thickness of the scattering material, up to a maximum thickness, which corresponds to one-third of the range of the particles in that material. The *back-scattering* factor is defined as the ratio of the measured intensity of a β source with a reflector, to that of the same source without a reflector. The maximum back-scattering factor obtainable with a sufficiently thick reflector is known as the *saturation back-scattering factor.*

Saturation back-scattering factors depend upon the atomic number of the scattering material, and may have a value as high as 1.75 (Fig. 4.5). That is to say, three quarters of the β-particles originally heading away from a detector will be reflected toward it. This in turn, implies that a β source prepared for counting on a metal of high atomic number, such as a platimum source planchet, could in fact give an apparent activity 1¾ of its true value. A similar source on a stainless steel planchet could give an apparent activity 1½ times its true value. While these errors are of little importance in relative counting, where the same type of planchets are used throughout, they are of considerable importance in absolute β assay. An additional complication in β assay is the production of bremsstrahlung. This may be produced as a result of interaction between a β-particle and a source planchet, or as a result of β interaction with the counter assembly. To reduce the production of bremsstrahlung, many counting assemblies are lined with a material of low atomic number *e.g.* aluminium or perspex.

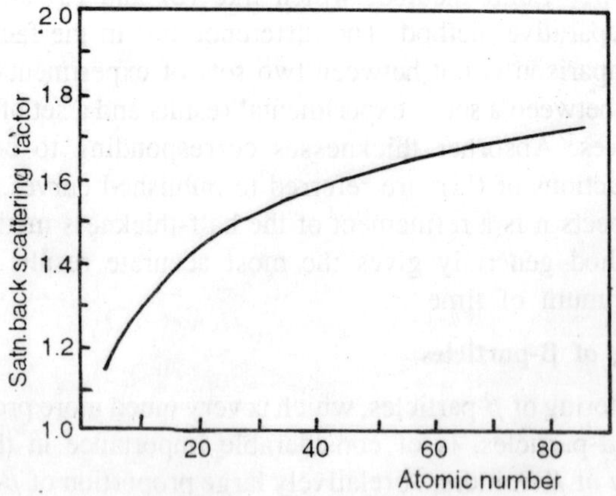

Fig. 4.5 : Saturation back-scattering factor as a function of atomic number of the scattering material.

Saturation back-scattering factors are independent of β-particle energy above 0.6 MeV. Below this figure the factors decrease with decreasing energy. The dependence on β energy is sufficiently significant below 0.3 MeV as to be used as the basis for a method for the determination of E_{max} for low energy β emitters.

POSITRONS (β^+)

The foregoing properties of energy spectrum range, etc. for β-particles are also true for positrons emitted during radioactive decay. Positrons have, however, one additional important property; they temporarily form a transitory positron-electron combination (positronium) after they have expended their kinetic energy. The coalition lasts for a period of 10^{-7} to 10^{-10} seconds, depending upon whether the particle "spins" are in the same or reverse directions. Thereafter the particles mutually annihilate each other with the simultaneous emission of energy in the form of γ-radiation. In the relatively long-lived case where the spins are in the same sense (ortho-positronium) three γ-rays with a total energy of 1.02 MeV are emitted and in the case where the spins are in opposite directions (parapositronium) two 0.51 MeV γ-rays travelling in opposite directions are emitted. The presence of 0.51 MeV γ-rays is frequently an indication of positron decay, and a 0.51 MeV peak is usually superimposed upon the γ spectrum of positron emitting nuclides.

γ-RADIATION

(A) Nature of γ-rays

Gamma-rays are a form of a very short wavelength electromagnetic radiation ($\lambda \sim 10^{-10}$ cm) and, like other forms of electromagnetic radiation, travel at the speed of light ($3 \cdot 10^{10}$ cm/sec). They are undeflected by electric and magnetic fields.

(B) γ Spectra

Gamma-rays are emitted with finite discrete energies (*i.e.* wavelengths) as a result of transitions between nuclear energy levels. Their spectra afford a valuable and highly developed method for the identification of radioactive species. Energies vary over a wide range but the majority lie between 0.1 and 2 MeV.

(C) Interaction with Matter. Mechanisms of Energy Loss

It is not possible to state an absolute value for the range of γ-rays because of the different competing mechanisms of energy loss, and the fact that ionisation is almost entirely secondary. For the same reason it is not possible to quote a figure for the specific ionisation of a single γ-ray. An average value for a large number of γ-rays would be from $^1/_{10}$ to $^1/_{100}$th that of a β-particle.

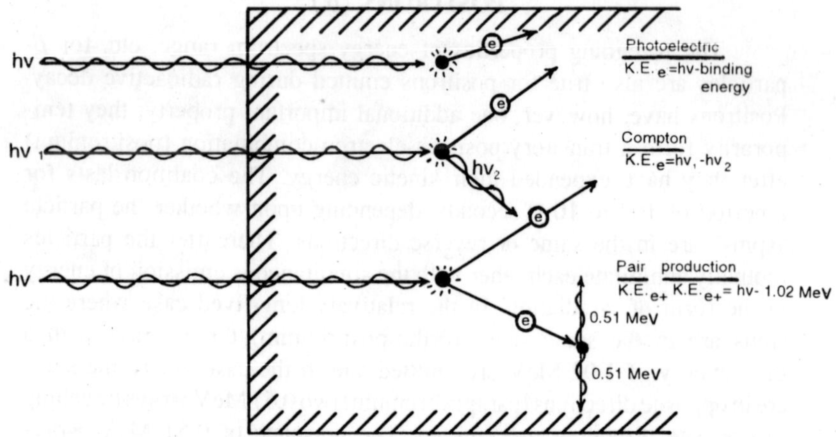

Fig. 4.6 : Diagramatic representation of the three mechanisms of γ energy loss.

There are three major mechanisms which cause γ energy loss (Fig. 4.6). These are :

(i) Photoelectric absorption, in which the γ-photon expends all of its energy to eject an orbital electron from an inner shell, with an energy equal to the original γ energy less the binding energy of the electron. The ejected electron thus has an energy comparable in magnitude to that of a β-particle, and is *directly related to the original γ energy.* The subsequent behaviour of the electron is similar to that of a β-particle.

(ii) Compton Effect, in which only *part* of the original γ energy is used to eject a bound electron. In this process the electron is from an outer shell where the binding energy is negligible compared with the initial γ energy. Thus, the sum of the energies of the scattered γ and the ejected electron may be taken as equal to the energy of the original γ-photon. The effect is named after compton, its discoverer, who also showed that the energy difference between the original and the scattered γ was mathematically related to the scattering angle. The

energies of the electrons ejected vary from zero to almost that of the original γ energy. Compton scattering is the predominant mode of γ energy loss for γ-radiation of about 1 MeV.

(iii) Pair Production is a unique process wherein energy is converted into mass. A γ-photon which has an energy greater than that equivalent to the rest mass of two electrons, *viz.* 1.02 MeV may, in the field of a nucleus, give rise to the creation of an electron-positron pair. In many respects this is the reverse of the electron–positron annihilation process. The probability of the process is greatly dependent upon γ energy, and becomes the most frequent mechanism for γ energy loss at energies above 5 MeV. The energy in excess of the threshold energy, appears as the kinetic energy of the two particles—there is no degraded γ. Both the positron and electron subsequently dissipate their energy in the manner of a β-particle. The positron, together with another electron is ultimately annihilated, with the appearance of the two characteristic 0.51 MeV γ-rays.

(D) Cross-sections of γ Absorption Processes

Cross-section (σ) was introduced in Chapter 3 to express the probability of a nuclear reaction being initiated by an incident particle. Total and partial cross-sections can also be attributed to atoms for γ interactions. The total corss-section σ_T, is equal to the arithmetic sum of the partial cross-sections σ_{pe}, σ_e, and σ_{pp}, for the three processes of photoelectric, Compton and pair production. Each of the partical cross-sections is a function of γ energy, and of the atomic number of the atom involved, as is shown in Fig. 4.7.

The *absorption co-efficient* μ for γ-rays is given by

$$\mu = N\sigma_T = N\,(\sigma_{pe} + \sigma_c + \sigma_{pp})\ \text{cm}^{-1}$$

where N is the number of atoms per cm³.

(E) Absorption of γ-radiation

The attenuation of γ-rays in a material is exponential; that is to say, if I_0 is the original intensity, then the intensity I after having passed through x cm of a material with an absorption coefficient of μ is given by :

$$I = I_0 e^{-\mu x}$$

or $$\log_{10} I/I_0 = -0.4343\ \mu x$$

from which it will be seen that a logarithmic plot of γ intensity against x should be a straight line.

The thickness of the absorbing material necessary to reduce the intensity to half of its original value is known as the *half-thickness* ($x_{½}$). It is related to the absorption coefficient μ by :

$$x_{½} = \frac{0.693}{\mu}$$

and the reduction in γ intensity after n half-thicknesses is $(½)^n$. For 1 MeV γ-rays in lead, the half-thickness is approximately 1 cm.

It is instructive to see the manner in which the three competing cross-sections (Fig. 4.7) affect the total cross-section for lead. It will be observed that lead is, for example, more "transparent" to γ-rays of the order of 5 MeV, than it is to those of either higher or of lower energy.

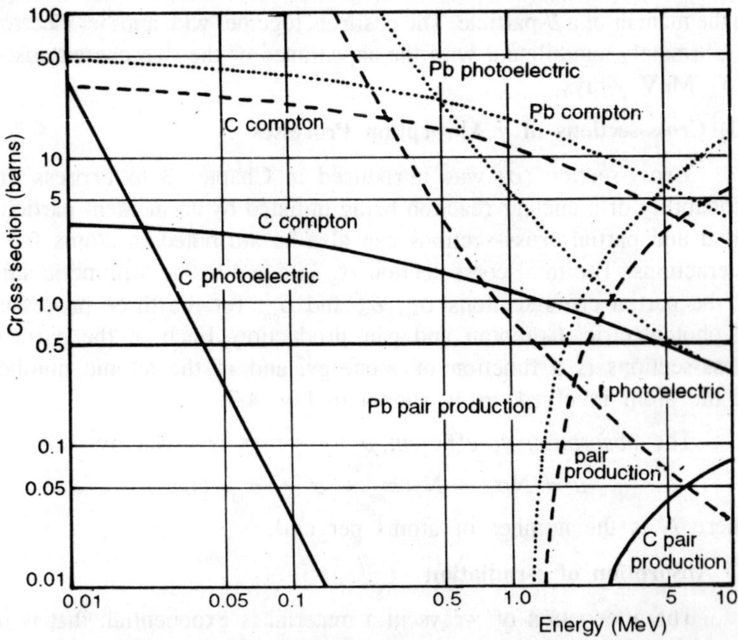

Fig. 4.7 : Cross-sections for the interaction of γ-radiation as a function of energy for various materials.

The analogous forms of the above equations and those describing radioactive decay, and half-life, will be apparent. Problems involving γ attenuation are also solved in an analogous manner—either arithmetically or graphically.

(F) Photo-nuclear Reactions

Besides the above mentioned atomic interactions of γ-radiation, there are rarer photo-nuclear reactions which occur when γ-radiation excites nuclear energy levels. The cross-sections are very low and do not materially efffect the total cross-section. However, some of the implications are important. Examples are the antimony–beryllium neutron source based on the γ, n reaction, and γ induced fission ih some heavy nuclei.

(G) Mossbauer Effect

A recently discovered phenomenon related to γ-ray nuclear excitation is the *Mossbauer effect.* This phenomenon, for which its discoverer received a Nobel prize in 1961, is likely to find increasing application by chemists as an analytical method and for the study of chemical bonds.

Previously in this book, it has been implied that a γ-ray is emitted with an energy exactly equal to the nuclear transition energy. If this were true, one would expect an absorber of the same isotopic material as gave rise to the γ-rays, and therefore with the same nuclear energy levels to exhibit resonant absorption (Fig. 4.8a). This is not normally observed. It is because it is not precisely correct to assume that the photon energy is exactly equal to the transition energy for two reasons:

(i) A small fraction of the transition energy (of the order of 0.1 eV) appears as recoil kinetic energy in the decaying nucleus. This is readily apparent when it is appreciated that, as a consequence of Einstein's relation, a γ-ray with energy E_γ has mass equivalent to

$$\frac{E_\gamma}{c^2}$$

and momentum

$$\frac{E_\gamma}{c}$$

Conservation of momentum requires the recoiling nucleus (mass m and velocity v) should have momentum equal to the γ momentum :

$$mv = \frac{E_\gamma}{c}$$

Thus the recoiling nucleus will have a velocity given by :

$$v = \frac{E_\gamma}{mc}$$

Now the recoil energy (E_R) = $E_\gamma^2/2mc^2$

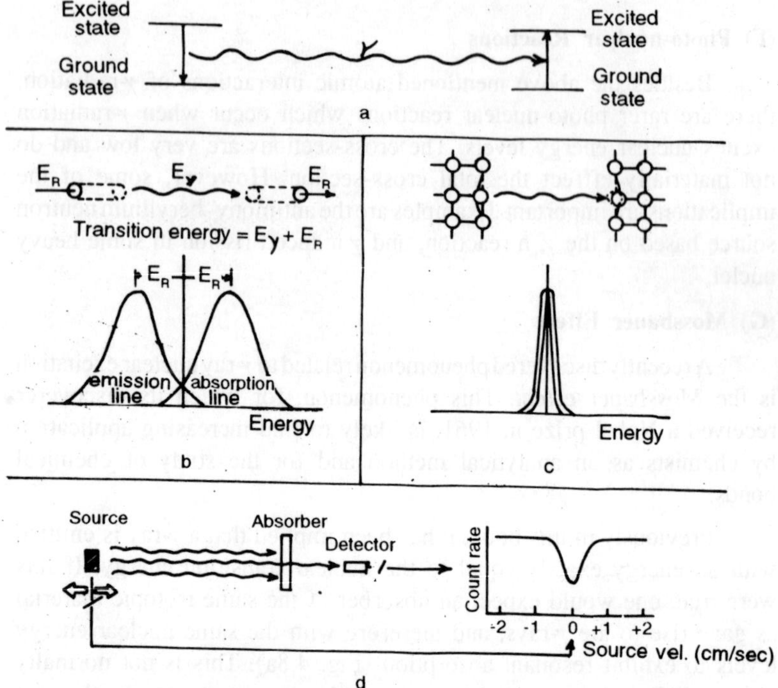

Fig. 4.8 : Diagramatic representation of the Mossbauer effect.

The term mc^2 is an expression of the energy equivalent of a mass m. If the mass m is expressed in a.m.u.—remembering that 1 a.m.u. is equivalent to 931 MeV then :

$$E_R = \frac{E_\gamma^2}{1862m}$$

where ER and Eγ are in MeV and m is in a.m.u.

It will be noted that the recoil energy is inversely proportional to the mass. The order of magnitude can be found by considering, as an example, 0.1 MeV γ-ray emitted from a free nucleus with a mass number of 50.

$$\text{Recoil energy} = \frac{(0\cdot 1)^2}{1862 \times 50}\text{MeV} = 0.1 \text{ eV}.$$

(i.e. the recoil energy is of the order of one millionth the value of the γ energy).

Approximate values of the recoil energy for various values of m and E_γ are shown in Table 4.2.

Table 4.2
Recoil Energy (eV) for Values of *m* and $E\gamma$

E_γ(MeV)	*m* = 5 a.m.u.	*m* = 50 a.m.u.	*m* = 500 a.m.u.
0.01	10^{-2}	10^{-3}	10^{-4}
0.1	1	10^{-1}	10^{-2}
1.0	10^{2}	10	1

Because of the reverse effect in the resonant absorber, a γ-photon would require an excess of energy equal to the same amount in order to be absorbed. The γ emission energy is therefore displaced from the absorption energy by twice the recoil energy (Fig. 4.8b).

(ii) The γ line width broadens as a result of the motion of the decaying atom. From the kinetic theory of gases, we know the atom has a velocity approximately 10^{-6} of the velocity of light at room temperature. The outcome of this is to increase, by the Doppler effect, the natural γ line width of less than a millionth of an eV, to a value of $10^{-6} \times E_\gamma$—or about 0.1 eV for a 0.1 MeV γ. Similar arguments apply to the absorbing nuclei.

The overall effect of (*i*) and (*ii*) above, is to produce a line broadening of about 0.1 eV for emission and absorption lines, and to shift them by about 0.1 eV from the true value of the transition energy. Thus there is no significant energy overlap to produce resonance absorption (Fig. 4.8b). Attempts in 1951 to observe resonant absorption, by moving source and absorber at high relative velocity, and produce a Doppler shift that would bring about overlap of emitting and absorbing energies, met with partial success.

Both recoil energy and Doppler effects can, however, be reduced to negligible proportions if both the radioactive nucleus and the absorbing nucleus are "anchored" in a crystal lattice (Fig. 4.8c). Recoil momentum is taken up by the crystal as a whole and since the effective mass is greatly increased, the recoil velocity and energy are reduced to negligible proportions.

Because crystals are elastic structures, recoil energy will appear in the vibrational energies of the crystal lattice. These energies are quantised. The probability that they will be excited depends upon their energy, the recoil energy available, and on the thermal excitation (crystal temperature). Providing the γ energy is of the order of 0.1 MeV or less, and the crystal temperature at, or preferably below, room temperature—there is a considerable probability, as shown by Mossbauer, of "recoiless" γ emission and absorption, with a line width of 10^{-6} eV or less. This corresponds *to a resolution of 1 part in 10^{-11}* and thus, resonant absorption and scattering are critically dependent on—and influenced by—very small changes in energy. If source and absorber are moved relative to one another with a velocity of only mm per sec. (10^{-11} the velocity of light) the Doppler effect is sufficient to remove the overlap and destroy the phenomenon of resonance absorption. A "resonance" spectrum can be scanned by simply moving a source and absorber relative to one another (Fig. 4.8d).

Experimental requirements are a low energy γ source, and absorber both incorporated in a crystal lattice, a radiation detector set to record only those γs having the resonant energy, and a means of producing controlled relative motion between source and absorber to scan the spectrum.

A chemical application is that of routine non-destructive analysis for certain elements; for example iron (^{57}Co source) and tin (^{119m}Sn source). Cross-sections for resonance absorption *at the resonant energy* are generally several hundred times greater than the next largest cross-section (photo-electric) and as a consequence, the effect can be observed when the nuclide concerned is only a minor constituent.

Changes in the nuclear environment will result in very slight changes in the energy levels of the emitting or absorbing nucleus, for example magnetic fields can produce a splitting of the nuclear energy levels which will consequently appear in a Mossbauer spectrum. A more important effect, as far as chemists are concerned, is the chemical shift which results from different electron densities near the nucleus. Although these energy shifts are only very slight, the extreme sensitivity of the method (1 part in 10^{11}) is adequate to provide information concerning chemical bonding in some crystal structures. Some chemical shifts for tin-halogen compounds are given in Table 4.3. In this latter application, the Mossbauer effect provides a method of inorganic structural study, somewhat analogous to the adaption of nuclear magnetic resonance to examine proton resonance shifts in organic systems.. The field of application is, however, somewhat more restricted.

(H) Szilard–Chalmers Reaction

In the previous section concerning the Mossbauer effect, we were concerned with the case where γ recoil energy was negligible. Let us now turn to the particular case of molecular species, where γ

Table 4.3

Bond	*Shift (mm/sec)*
Sn–Sn	0.0
Sn–I	0.3
Sn–Br	1.0
Sn–Cl	1.3
Sn–F	2.6

recoil energy is present as an aftermath of neutron irradiation, and is greater than the bond energies. This will be the case for *g* transitions of greater than about 1 MeV (Table 4.2).

These are the conditions necessary for a reaction, originally discovered by *Szilard and Chalmers* in 1934. permitting the seemingly impossible chemical isolation of a radioactive isotope ^{128}I from an irradiated target material containing stable isotopic ^{127}I. As mentioned previously, neutron irradiation of a target produces an isotope of that target material having a mass number one greater. The energy available from the binding energy of the absorbed neutron, leaves the nucleus in an excited energy state and is subsequently emitted as a γ-ray.

If the recoil energy produced is in excess of bond energies (2 to 5eV), bond rupture generally occurs. Note however, that bond rupture is restricted to those molecules containing the nuclei which have undergone the n, γ reaction; *i.e.* those which have formed the new isotopic species. The isotopes thus produced are now likely to be found in a chemical form distinctly different from the compound irradiated, and are therefore capable of chemical isolation, providing there has been no subsequent chemical recombination, or that no exchange reaction has occurred with the target compound. Most Szilard–Chalmers reactions are based upon neutron irradiation of a covalent target compound and subsequent isolation of elementary or ionic products.

(I) Determination of γ energy — γ spectrometry

An early method for the determination of γ energy was the utilisation of the Compton effect with a suitable "radiator" such as copper, used in conjunction with a γ source to provide Compton electrons of varying energies. Techniques, similar to those for determining β_{Emax} energy. This method is now superceded by the scintillation γ spectrometer. This instrument converts light flashes, produced in a phosphor by γ-radiation, into electrical pulses of a size proportional to the γ energy. These pulses are then sorted out electronically.

5

The Detection and Measurement of Nuclear Radiation

An instrument for the detection and measurement of radiation generally consists of two parts : a detector which converts the energy of the radiation into an electrical pulse, and an electronic circuit which amplifies and records the pulses so produced.

In this Chapter the principle of operation of types of detector will be surveyed, and the auxiliary electronic equipment will be represented by a block diagram. Some of the basic functions of this electronic equipment will be further described in Chapter 6.

Detectors at present in use may, for convenience, be divided into three classes :

(1) Gas ionisation detectors.
(2) Scintillation detectors.
(3) Solid state detectors.

Gas Ionisation Detectors

A gas is normally a non-conducting medium because there are no charge carriers present. However, the passage of radiation can produce ion pairs about one for every 30 eV of energy expended by the radiation. If these ion pairs (electrons and positive ions) are collected by two electrodes before recombination can occur, a small electrical pulse will be observed.

Methods based on gas ionisation may be subdivided according to the electric field gradients employed. If we imagine a detector consisting of two electrodes in a gas, which is ionised by the passage of radiation, and if we graph the size of the electrical pulses as a function of the potential applied between the electrodes, we would obtain a curve similar to that shown in Fig. 5.1.

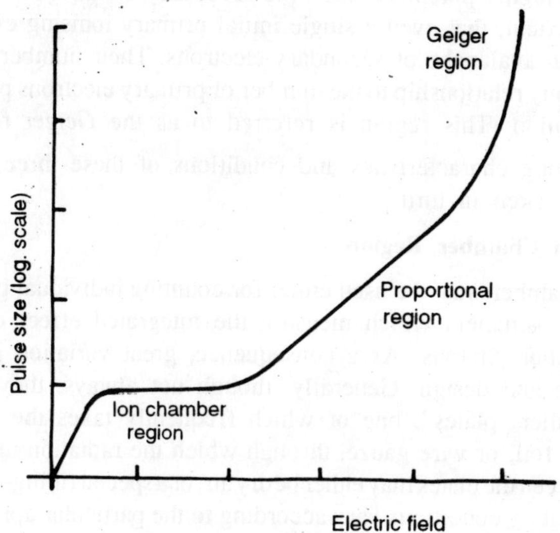

Fig. 5.1 : Variation of pulse size as a function of applied potential for gas ionisation detectors

Without any potential difference to separate and collect the ion pairs there would, of course, be no pulse. As the potential is increased, more and more of the elctrons and positive ions produced by the passage of the radiation will be attracted to and collected by the electrodes until a situation is reached where the potential is large enough to collect all elctrons and positive ions produced by the passage of the radiation. A further small increase in potential will not produce any greater pulse. This region of operation is referred to as the *ion chamber* region.

If the potential between the electrodes is further increased, the electrons produced by the ionising radiation will be accelerated to a velocity—and therefore energy—sufficient to produce secondary ionisation by collision with the atoms or molecules of the gas. Some of these electrons produced by secondary ionisation will, themselves, acquire sufficient energy from the field to produce further ionisation. While the pulse collected will be considerably larger than would have resulted from the collection of the primary electrons only, it will, nevertheless, still be proportional to the number of original primary electrons produced by the radiation. This region is the *proportional* region.

At yet higher potentials, multiple secondary ionisation will occur to such an extent, that even a single initial primary ionising event can trigger off an avalanche of secondary electrons. Their number will no longer bear any relationship to the number of primary electrons produced by the radiation. This region is referred to as the *Geiger* region.

Operating characteristics and conditions of these three regions will not be taken in turn.

(A) The Ion Chamber Region

Ion chambers may be used either for counting individual particles, or as β, γ dosimeters which measure the integrated effect of many particles and/or photons. As a consequence, great variation is found in their size and design. Generally, though not always, they consist of two parallel "plates", one of which frequently takes the form of a thin metal foil, or wire gauze, through which the radiation may pass. The gas between the plates may either be dry air, or a special filling—usually argon. Operating conditions vary according to the particular application. The pressure is frequently atmospheric, and potentials which must be sufficiently high to sweep out the ion pairs as rapidly as possible, yet not high enough to cause secondary ionisation, vary from 50 to 500 volts depending upon the design of the chamber.

Only individual α-particles (and fission product fragments) can produce sufficient ionisation to be detected as individual events, as the following example illustrates :

Consider an α-particle of 5 MeV. It will have a range at atmospheric pressure comparable to ion chamber dimensions, and so we will assume that all of the 5 MeV energy is expended within the chamber.

$$\text{Number of ion pairs} = \frac{5 \times 10^6}{30}$$

$$\text{Net charge collected} = \frac{5 \times 10^6}{30} \times 1.6 \times 10^{-19} \text{ coulombs (electronic}$$

charge = 1.6×10^{-19} coulombs)

$= 2 \times 10^{-14}$ coulombs (approx.)

Assuming a typical value of 10^{-11} farad, for the electrical capacity of the chamber then :

$$\text{Voltage pulse} = \frac{\text{charge}}{\text{capacity}} = \frac{2 \times 10^{-14}}{10^{-11}} = 2 \times 10^{-3}$$

With amplifier gains of 10^4 a two millivolt pulse is capable of amplification to a value when it will operate a scaler or ratemeter (Fig. 5.2a). In addition, for a chamber large enough for the α-particle to dissipate all of its energy, the pulse size is a function of the initial α-particle energy, and with the aid of a pulse height analyser, the system may be used to measure α-particle energies.

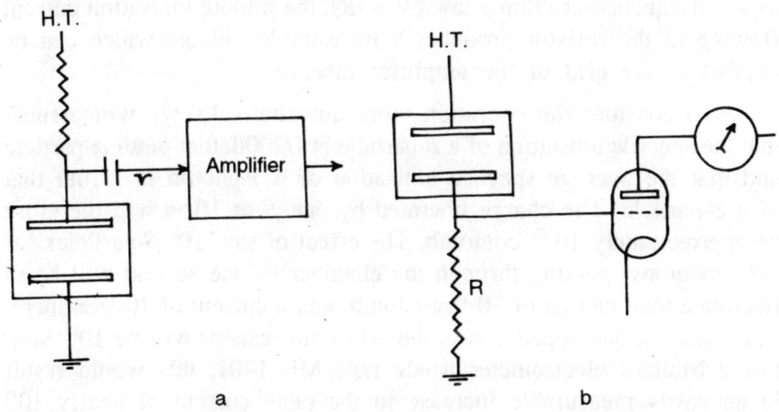

Fig. 5.2 : Simplified circuit for the operation of an ion chamber.

(a) To detect individual particles.

(b) To measure the integrated effect of a large number of particles and/or γ-photons.

In a neutron sensitive modification of the ion chamber, an internal deposit of a fissionable nuclide is used. The presence of neutrons is detected by the ejection of heavy charged, fission fragments into the volume of the counter.

Because the specific ionisation of β-particles is only 1/200th to 1/500th that of α-particles, the voltage pulse obtained from them is correspondingly smaller. It is, in fact, less than amplifier "noise" levels. Individual β-particles cannot, therefore, be detected by using an ion chamber.

The integrated effect of many β-particles (and also γ-photons) can, however, be measured by incorporating the ion chamber in a circuit with a large series resistor. If the time constant of the circuit (capacity × resistance) is far greater than the time interval between the arrival of consecutive particles or photons, a steady ionisation current is produced and can be measured by a simple d.c. amplifier circuit. Such an ionisation device is the basis of many portable laboratory β, γ radiation survey meters. The important feature of these instruments is the very high value of resistance—frequently of the order of 10^{10} ohms—which is placed in series with the ion chamber (Fig. 5.2b). This resistor plays two important roles. Firstly, it provides a circuit with a long time constant, and thereby integrates the pulses into a d.c. ionisation current. Secondly, as a consequence of Ohm's law ($V = iR$), the minute ionisation current flowing in the resistor produces a measurable voltage which can be applied to the grid of the amplifier tube.

To consider the operation more quantitatively, we will assume that the specific ionisation of a β-particle is 1/200th that of an α-particle and that the average specific ionisation of a γ-photon is 1/10th that of a β-particle. The charge liberated by one β or 10 γs will therefore be approximately 10^{-16} coulomb. The effect of say, 10^5 β-particles (or 10^6 γ-photons) passing through the chamber in one second will be to liberate a total charge of 10^{-11} coulomb; i.e., a current of 10^{-11} ampere. The potential developed across the 10^{10} ohm resistor will be 10^{-1} volt. For a Mullard electrometer triode type ME 1401, this would result in an easily measurable increase in the plate current of nearly 100 μA—a current amplification of 10^7.

One unusual variety of ionisation chamber which deserves particular mention is the gold leaf electroscope, used by many of the early research workers prior to the advent of electronics. The electroscope was charged, and the rate of collapse of the leaves, which was a function of the degree of ionisation—and hence activity—was measured. A modern counterpart of this instrument is to pocket dosimeter, which is about the same size and shape as a fountain pen. After being charged from a power pack, it is worn, and the accumulated radiation dose is measured by the drift of a quartz fibre, as seen against a calibrated screen, when the dosimeter is held towards a source of light.

(B) The Proportional Region

At higher potential gradients, electrons produced by the ionising radiation and subsequently accelerated by the electric field, produce secondary ionisation by collision with neutral gas atoms and molecules. The degree to which this occurs is a function of the design and the operating potential, and is called the "gas amplification factor". It is usually whithin the range 10^2 to 10^4. That is to say, for every electron produced by the radiation, 100 to 10,000 are collected at the anode. The total number collected is, however, still proportional to the original number produced by the radiation.

The higher potential gradients necessary to bring about the effect are obtained—not by applying higher potentials between two plates—but by making the anode in the form of a thin wire within a cylindrical or hemisperical cathode. This produces a non-uniform potential gradient. In such an arrangement, illustrated in Fig. 5.3, the potential gradient

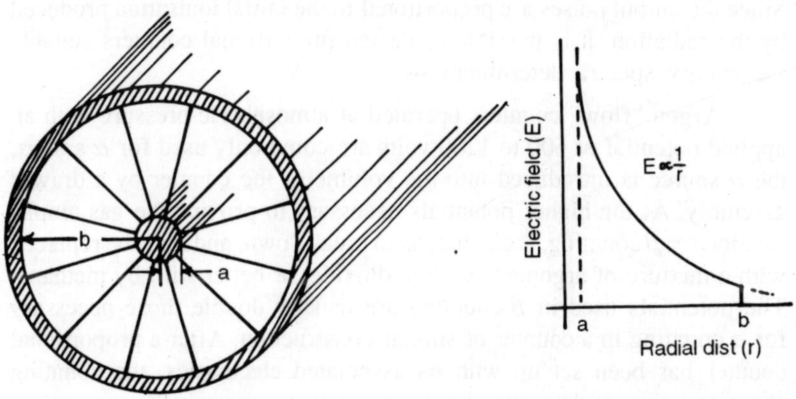

Fig. 5.3 : Production of a non-uniform potential gradient in a proportional counter.

E along the radius $\left(\frac{dV}{dr}\right)$ is, by definition, proportional to the density of the lines of electrical force and therefore :

$$E \propto \frac{1}{r}$$

or

$$E = \frac{k}{r}$$

Substituting $\frac{dV}{dr}$ for E and integrating gives :

$$V = k \log_e \left(\frac{b}{a} \right)$$

from which the value of k, and hence E, at any point can be determined if the applied potential (V), the radius of the wire (a) and the radius of the cathode (b) are known.

In a counter of one inch diameter and with a 0.001 inch diameter anode wire, an overall applied potential of 1,000 volts would result in a potential gradient varying from about 10^2 V/cm at the cathode wall to about 10^5 V/cm at the anode. It is in this region, very close to the anode wire, that secondary ionisation occurs.

The proportional counter is frequently used for α assay, and because of the "gas amplification" it is also possible to use it for β counting. Since the output pulses are proportional to the initial ionisation produced by the radiation, it is possible to design proportional counters suitable for energy spectra determinations.

Argon "flow" counters operated at atmospheric pressure with an applied potential of 600 to 1200 volts are commonly used for α assays; the α source is introduced into the volume of the counter by a drawer assembly. At the higher potentials necessary to provide the gas amplification for β counting, argon begins to break down, and must be replaced with a mixture of argon and carbon dioxide, or better still, by methane. The potentials used in β counting are usually double those necessary for α counting in a counter of similar construction. After a proportional counter has been set up with its associated electronics, the counting characteristics are determined experimentally to ascertain the appropriate operating conditions.

One particular application of β proportional counting that deserves special mention is the carbon dioxide filled counters used in carbon dating. the filling gas which is frequently obtained by the controlled combustion of some archeological sample, performs the dual function of both source and counting medium. The radioactive carbon-14 content of the gas emits β-particles within the counter and produces initial ionisation. Electrons produced by this ionisation are accelerated by the field to produce secondary ionisation in the carbon dioxide near the wire anode. Methane is also used in place of carbon dioxide in carbon dating counters. Although the gas preparation is more complex it has the advantage of being less affected by electronegative impurities.

Another variation in proportional counting is the boron trifluoride filled counting tubes used for neutron detection. Neutrons themselves produce no ionisation, but react with boron, which has a high capture cross-section, causing an α-particle to be emitted by the ^{10}B (n, α) ^{7}Li reaction. In this case the gas acts as both the counting medium and the source of ionising radiation.

(C) Geiger Region

As pointed out at the beginning of this Chapter, higher potential gradients can trigger off an avalanche of secondary electrons the magnitude of which bears no relation to the original number of primary electrons produced by the radiation. Just as the conditions necessary to effect proportional counting characteristics were brought about by a change in geometry, rather than by an increase in applied potential, operation in the Geiger region is achieved by yet another change in design. This time the geometrical property of a wire anode is retained, but the gas pressure is reduced to about one tenth of an atmosphere. This results in an increase in the mean free path of the electrons which are accelerated by the applied field, and operation in the Geiger region can be achieved with potentials of the same order as those previously used in a proportional counter at atmospheric pressure.

In many ways, a Geiger tube (or Geiger–Müller tube as it is sometimes called) can be thought of as operating in a state of metastable equilibrium. Potential gradients are so high, that once a Geiger tube has "fired", it would continue in a state of continuous discharge were it not for some form of internal or external quenching. This is due to the positive ions which are formed near the anode moving relatively slowly to the cathode, where they combine with an electron. With an ionisation potential for argon of 15.7 eV a considerable amount of energy is available either to liberate an electron from the cathode, or to emit a photon which in turn liberates a photoelectron elsewhere in the tube. These electrons then begin a second avalanche with the production of another batch of positive ions, and so the process is repeated.

The initial discharge can be quenched, however, by either internal chemical means, or external electronic means. In practice, both are often used.

Chemical quenching is achieved by the addition of a polyatomic vapour to the counting gas to sequester the positive ions produced in the initial discharge. The ionised gas deionises at the expense of ionising the polyatomic molecule which has a much lower ionisation potential. This in turn deionises and uses up the surplus energy by dissociating into fragments. Older, organically quenched tubes contain about 90% argon and 10% ethyl alcohol or ethyl formate vapour. In each discharge, some 10^{10} molecules are consumed and the total counting life of such tubes is about 10^{10} counts. More recently developed halogen quenched tubes employ some 0.1% chlorine or bromine, which after dissociation, reconstitutes itself and so these tubes have, theoretically, an infinite life.

Electronic quenching terminates the discharge by lowering the potential allied to the tube to a potential below the Geiger region after the tube has fired. This is most simply accomplished by applying the high tension to the tube through a large value resistor (Fig. 5.4), so that the time taken to return the anode to its operating potential after initial discharge is long, compared with the time taken for the positive ions to move to the cathode which is normally of the order of 10^{-4} seconds. With a Geiger tube electrical capacity of 10^{-11} farad this would entail the use of a load resistor of some 10^8 ohms ($C \times R = 10^{-3}$ sec). The main difficulty with this simple method is that recovery is exponential with resulting uncertainty about the exact "dead-time" of the tube. A better solution—and one which also serves to provide a

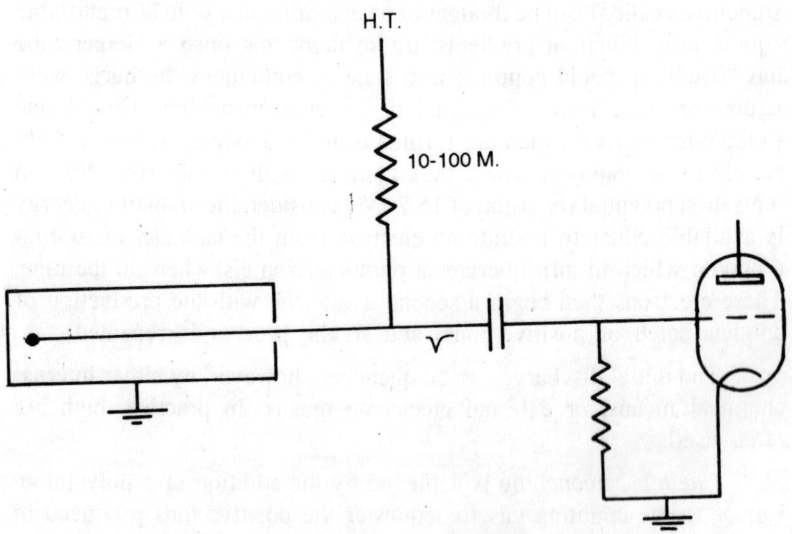

Fig. 5.4 : Simplified circuit for the operation of a Geiger detector.

known fixed "dead-time" to permit coincidence losses to be computed—is that of superimposing a large negative square wave pulse on the tube supply. The significance of coincidence losses and dead-time are treated later in Chapter 11. The determination of dead-time is given in Expt. 5.

The serve a large number of different applications, Geiger tubes are made in a great variety of shapes and sizes. Some of the more common types are illustrated in Fig. 5.5. The operating characteristics depend on the type of tube. Threshold potentials vary from 300 to 1200 volts, output pulses from 1 to 50 volts amplitude, while dead-times of the order of 200 μsec are common. Lower threshold potentials and higher output pulses are usually associated with the halogen quenched tubes. Because of the large output pulse magnitude, little or no amplification is required to operate a scaler or ratemcter. Furthermore, as the operating plateaux are relatively long and flat, stabilised high tension supplies are not essential, and the electronics associated with Geiger detection are relatively simple and cheap. The method for the determination of the plateau characteristics is given in Expt. 3.

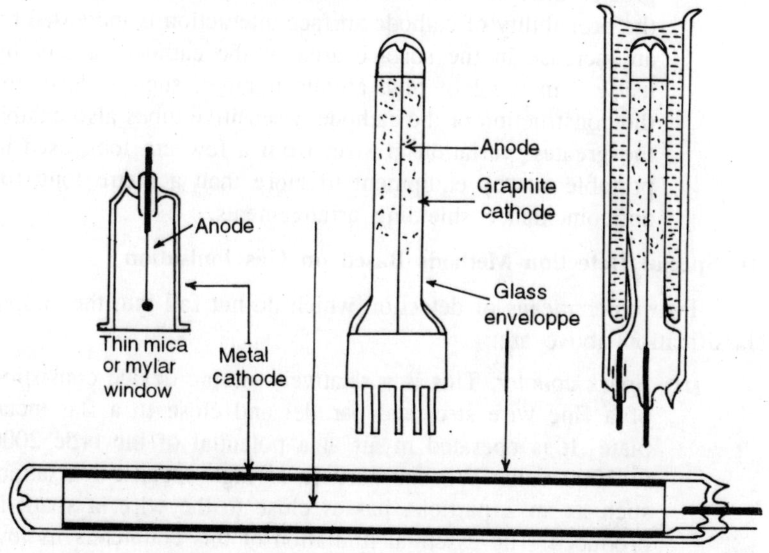

Fig. 5.5 : Types of Geiger detector in common use, thin end window, thin walled tubular for β detection, liquid counting type and metal cathode variety used for higher efficiency γ-ray detection.

Types of Geiger tubes used to detect the different types of radiation are :

(i) *α detection.* Thin end window types with a window thickness less than 2 mg/cm^2 will detect α-particles with 100% efficiency.

(ii) *β detection.* End window, thin glass walled tubular, and liquid counting types may all be employed. Except for the lowest energy βs, counting efficiencies in the thin end window types will approach 100%. At the other extreme, losses in penetrating the wall and in the solution media, will reduce the counting efficiency of liquid counting types to between 2% and 4% for 0.5 MeV βs. See Expt. 12.

(iii) *γ detection.* The majority of γ-photons will pass right through a Geiger tube without producing a single ionising event. Counting efficiency for γ detection, depends, not on penetration, but on the probability of an interaction resulting in the emission of an electron—usually from the cathode—into the active volume of the counter. γ detection efficiencies are seldom greater than 1%. To raise this tubes are available wherein the probability of cathode surface interaction is increased by an increase in the surface area of the cathode, and/or by using a material of high atomic number, such as lead, for the construction of the cathode. γ sensitive tubes also exhibit the greatest variation in size; from a few cm long used in portable survey equipment to more than a metre long for anticoincidence shielding arrangements.

D) Special Detection Methods Based on Gas Ionisation

Two other means of detection which do not fall into the simple classification above are:

(i) *Spark counter.* This is a relatively simple device consisting of a fine wire stretched parallel and close to a flat metal plate. It is operated in air at a potential of the orde 2000 to 3000 volts. When a particle of high specific ionisation, such as an α-particle, passes close to the wire, a spark is produced. The essential limitation of this counter is its low geometric efficiency.

(ii) *Cloud Chamber.* The cloud chamber is employed not for counting particles, but for studying their properties by direct visual, or by photographic means. There are two types of cloud chamber, but the principle in each is the same—condensation from a saturated atmosphere around ions formed along the path of ionising radiation. The two types are : the Wilson cloud chamber, which provides the necessary conditions over a short period of time, by the sudden, adiabatic expansion of moisture laden air; and the diffusion cloud chamber, which provides a vapour laden atmosphere, by the slow distillation of a suitable volatile compound (*e.g.* alcohol), across a temperature gradient.

The Scintillation Counter

The scintillation counter represents an electronic development of the early spinthariscope, a device which consisted of a zinc sulphide scintillation screen viewed by a lens, and which was used extensively by Rutherford and his contemporaries for α-particle detection. The tedious visual counting of light flashes which was necessary, resulted in the 1920's and 1930's in the adoption of gas ionisation detectors which were readily adaptable to electronic recording. However, with the development of sensitive photomultiplier tubes which could "see" a light flash, and produce a measurable electrical pulse, there was a general revival of scintillation techniques after the second World War and during the 1950's. With the development of a wide range of scintillating media, or *phosphors*, as they are called, the method is adaptable, not only to α, β and γ detection, but also to the measurement of their energy spectra. Because of their reliability, versatility, and rapid resolving time, scintillation techniques are very widely used at present.

The basic components of a scintillation counter are shown diagramatically in Fig. 5.6. A source of stabilised high potential operates the photomultiplier which "sees" the light flash produced in the phosphor

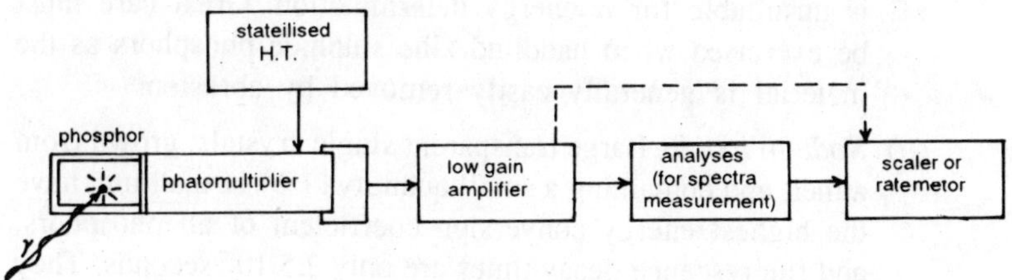

Fig. 5.6 : Typical arrangement for scintillation counting.

by the incident radiation. The small electrical pulse from the photomultiplier is amplified, and either passed directly to a scaler or ratemeter if gross counting is required, or via a pulse height analyser to a recording device, if energy spectra measurements are to be made.

The operation of each of these parts is described below :

(A) The Phosphor

Although called a phosphor, the phenomenon observed is not phosphorescence, but fluorescence, in nearly all cases, and results from electronic excitation produced by the incoming radiation. The duration of the light flash is of the order of 10^{-6} sec for inorganic and 10^{-8} sec for organic phosphors. With suitable electronics, extremely rapid resolving times are possible. So fast is the response time of the detector, that frequently the upper limit to counting speed is set, not by the detector, but by the associated electronic scalers. Various types of phosphor are now used; zinc sulphide is still employed for α counting, but, in addition, other inorganic crystals, organic crystals, plastics and solutions are used for a variety of purposes. The greater density of these materials, compared with gases, favours energy dissipation. This results in higher detection efficiencies compared with gas ionisation detectors—especially in the case of γ-photon detection. Phosphors which are in common use, and commercially available are :

(i) *Zinc Sulphide* as a semi-transparent deposit on a glass flat. The deposit must be thin enough for scintillations produced by α-particles impinging on one side to be visible to the photomultiplier on the other. As a result, it is too thin to dissipate enough energy from α- or γ-radiation for these to be detected. The phosphor is, therefore, ideally suited to low background α counters. Becasue all of the α energy does not produce a flash visible to the photomultiplier, this phosphor is unsuitable for α energy determination. Great care must be exercised when handlind zinc sulphide phosphors as the material is generally easily removed by abrasion.

(ii) *Sodium iodide.* Large transparent single crystals, grown from a melt and containing a small quantity (1%) of thallium, have the highest energy conversion coefficient of all phosphors, and fluorescence decay times are only $2.5 \cdot 10^{-7}$ seconds. They have a high density and are largely composed of an atomic species (iodine) with a high atomic number, and therefore have a large cross-section for γ-ray absorption (Fig. 4.7).

Unfortunately, the material is deliquescent and must be encapsulated in a thin metal can with a glass window. The detection of α-particles and soft β-particles is therefore not possible since they cannot penetrate the container.

(iii) *Caesium iodide*, also thallium activated, can be grown as large transparent crystals. It has a lower energy conversion coefficient (about $^2/_3$ that of sodium iodide), a much longer light decay time ~ 10^{-5} sec, but it is not deliquescent. Since it does not need to be encapsulated, it is suitable for all types of radiation, including α scintillation spectrometry. Its high cost relative to other types of phosphors, its long decay time, and the advent of solid state detectors at about the same time as its commercial production, have largely restricted the adoption of this phosphor.

(iv) *Organic crystals.* Many cyclic organic compounds can be employed. One of the most common is anthracene, which has the highest energy conversion of the organic phosphors, with a value of half that of sodium iodide, and a fluorescence decay time of less than 10^{-8} seconds. Some disadvantages of anthracene, and other organic crystals, are their high cost, their low atomic number composition with correspondingly low photoelectric absorption efficiency for γ-photons (Fig. 4.7), and deterioration with consequent with consequent loss in light transmission.

(v) *Liquid phosphors.* As the term implies, these phosphors are liquid solutions which contain a fluorescent compound; for example, anthracene, or *p*-terphenyl, in a suitable solvent such as xylene. A 5g/litre solution of *p*-terphenyl solution in xylene has approximately one half the energy conversion of an anthracene crystal. A similar solution of anthracene has only one-tenth the efficiency of a *p*-terphenyl solution. Such solutions are placed in a transparent receptacle—frequently a glass beaker—on top of a photomultiplier. The assembly is then covered with a light tight shield. Liquid phosphors are cheap, effective, and particularly useful in measuring activity arising from organic soluble compounds which are dissolved in the solvent along with the fluorescent compound. This technique is particularly efficient for carbon-14 and tritium labelled organic compounds, which are often difficult to measure by other methods due to the low β energies.

(vi) *Plastic phosphors.* Fluorescent compounds such as anthracene, can be incorporated as a solid solution in several transparent plastics. The most commonly used plastic is polystyrene. A 2% anthracene solution in polystyrene is 20% as efficient as pure anthracene. These phosphors have the advantages that they are cheap, robust, can be prepared in any size or shape, machined to suit a particular application, and do not need to be encapsulated. They are suitable for α, β, and γ detection. One disadvantage in the case of γ detection, is their composition from elements of low atomic number. This results in a low photo-electric absorption cross-section (Fig. 4.7), thus making them unsatisfactory for γ energy determination, and inefficient for low energy γ detection.

Boron loaded plastic phosphors are also available for neutron detection and measurement.

The relative merits of the various phosphors available in relation to the type of radiation to be detected and the technique employed are summarised below.

(i) *α detection.* The normal procedure is to slide the a source under the phosphor via a light tight drawer assembly. Phosphors employed are :

(a) Zinc sulphide, where exclusive α detection is required in the presence of other types of radiation, or where a low background is necessary. Background counts as low as a few counts per hour are possible.

(b) Plastic phosphors where use of the same piece of equipment is required for β detection, and where low backgrounds are not important.

(c) Caesium iodide for a enrgy determination by scintillation techniques.

(ii) *β detection.* A drawer assembly can be used, and is preferable for low energy β-particles from a solid source. Alternatively, a light tight encapsulated phosphor, either in the shape of a cylinder, or a "well" (for solution counting) can be used. Because β sensitive phosphors are also γ sensitive, shielding is generally required to reduce the background count rate. Suitable phosphors are :

a) Plastic (or anthracene) especially where lower γ efficiency is desired. Plastic phosphors are preferable due to their much lower cost.

b) Sodium iodide where the β-particles possess sufficient energy to penetrate the aluminium can, and particularly where both β and γ detection are required in the one instrument.

c) Liquid scintillation techniques—these are more especially suitable for carbon-14 and tritium compounds.

(iii) γ *detection.* Detection assemblies take many forms. They vary from heavily shielded "fixed" installations, to unshielded probes consisting of a crystal, photomultiplier and small preamplifier arranged axially in the same light tight can. Suitable phosphors are :

a) Sodium iodide which is preferable because of its high atomic number composition, and therefore high γ cross-section (Fig. 4.7).

b) Plastics, though less efficient as a detection medium, find many general applications because of their lower cost. They also find some special applications in fast counting work, because of the more rapid fluorescence decay time.

(iv) γ *spectrometry* by scintillation techniques deserves particular mention. It is of great value, both in the identification of a radioactive species, and for measuring the activity of a particular γ emitter in the presence of a large γ background. The basic requirement of the detector for this application is that it should produce a light flash proportional to the γ energy; *i.e.*, the γ-ray should dissipate all of its energy within the phosphor. This is fulfilled if the γ-photon undergoes photo-electric absorption, but not if the γ-photon undergoes Compton absorption—unless the degraded γ-ray undergoes subsequent photoelectric absorption.

The experimental aim, therefore, is to promote photoelectric absorption, and either to inhibit or to compensate for Compton absorption. This is done by :

a) Promoting photoelectric absorption using a phosphor with a high photo-electric cross-section; *i.e.* one containing elements of high atomic number such as NaI (Fig. 5.7), preferably using a large crystal to increase the probability of any degraded γ undergoing photo-electric absorption before it leaves the crystal.

b) Compensate for Compton absorption. There are two methods by which this can be achieved.

One is to surround the sodium iodide crystal with a ring of γ detectors to detect Compton scattered γ-rays and to trigger an anti-coincident circuit to cancel out the electrical pulses from the scintillation detector. A difficulty with this method lies in the fact that only a proportion of the degraded Compton γ-photons are detected.

A second method uses two scintillation detectors, one with a sodium iodide crystal, and the other with a plastic phosphor. Reference to Fig. 5.7 shows that both detectors have a significant probability for Compton absorption. By feeding the integrated outputs from both detectors in opposite polarity into the same recorder, the Compton effect produced in the sodium iodide can be subtracted. This is known as the double crystal γ spectrometer.

(B) The Photomultiplier

A photomultiplier consists of a photosensitive cathode, which emits electrons when the light from the phosphor falls upon it. The light intensity from the phosphor is, however, so low that the number of electrons ejected from the photocathode could not be measured by conventional electronic means. These electrons are therefore accelerated within the photomultiplier tube by some 100 volts toward a "*dynode*", where each photoelectron gives rise to an average of approximately four secondary electrons. These electrons are subsequently accelerated toward a second dynode, held at a potential some 100 vots higher than the first, where each gives rise to another four and so on. If, x, is the number of secondary electrons produced per bombarding electron at each dynode, then the electron multiplication for one stage is x, for two stages it is x^2 and for n stages it is x^n. Most commercial tubes have from 9 to 14 stages. A ten stage tube, assuming $x = 4$, would have an electron multiplication of 4^{10}, or in other words, about one million.

The two most common dynode arrangements used are the *focussed electrode* system (Fig. 5.7a), which employs interdynode potentials of some 100 volts, and the *venetian blind* system (Fig. 5.7b) which employs potentials some 50% higher. Interstage potentials are derived from a

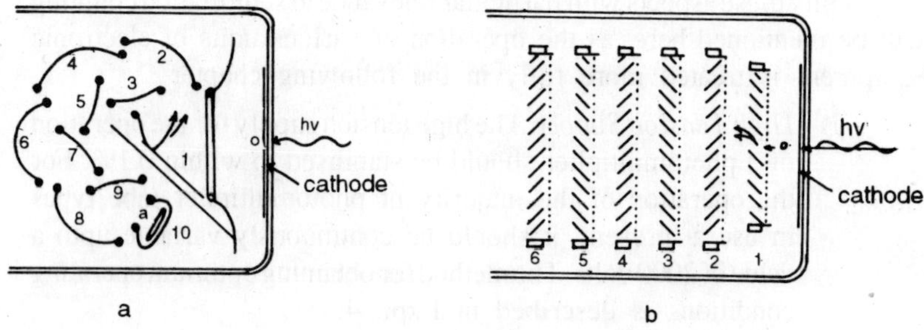

Fig. 5.7 : Common types of photomultiplier :

a) the focussed electrode system;

b) the venetial blind electrode system.

single high tension supply, by a voltage dividing resistor network—or "dynode chain"—on the photomultiplier tube base. Because the tube gain is critically dependent on inter-dynode potentials, the high tension supply should be stabilised to ±0.1%, and high stability resistors should be employed in the dynode chain.

The need for the total exclusion of external light from a photomultiplier assembly is obvious, but not so obvious is the need to exclude external magnetic fields, which by their influence on electron trajectories—particularly in the first stage—seriously impair tube performances. This applies to such weak fields as the earth's magnetic field (0.8 Orested), which in some cases can reduce the tube gain by as much as two thirds. Tube noise resulting from thermal electron emişsion from the cathode can, in certain applications, be troublesome. In such cases it is normally overcome by refrigeration of the photomultiplier tube assembly.

Some of the charactertistics of a few common tube types are given in Table 5.1.

Table 5.1

Tube type		*Dynode construction*	*No. of stages*	*Overall gain*
RCA	931A	Focussing	9	10^6
RCA	5819	Focussing	10	$6 \cdot 10^5$
RCA	8575	Focussing	12	$4 \cdot 10^6$
EMI	5060	Venetian blind	11	10^7
Philips	56AVP	Focussing	14	10^8

(C) Auxilliary Electronic Equipment

Only those aspects with particular relevance to scintillation counting will be mentioned here, as the operation of various units of electronic equipment is treated more fully in the following chapter.

(i) High Tension Supply. The high tension supply for the operation of a photomultiplier should be stabilised to within 0.1%. For the operation of the majority of photomultiplier tube types in use at present, it should be continuously variable upto a value of 2000 volts. The method for obtaining optimum operating conditions is described in Expt. 4.

(ii) Amplifier. For the majority of tubes in common use a low gain amplifier is require before the pulse is fed to the scaler or ratemeter.

(iii) Analyser. If energy spectra are to be determined a pulse height analyser is employed between the amplifier and the recording instrument.

(iv) Recording instrument. For normal counting work, this takes the form of either a scaler, for registering the gross counts in a fixed time interval, or a *ratemeter* to provide a continuous record of the rate at which the pulses arrive. For spectra determinations, the recording device may take the form of a chart recorder coupled to a single channel analyser, a bank of scalers used in conjunction with a multichannel analyser, or a modern transistorised multichannel analyser capable of analysing the spectrum into several hundred channels, each capable of storing 10^6 counts in a magnetic core memory. The spectrum thus analysed may be displayed on a C.R.O. screen, and X–Y recorder, punched out automatically on a typewriter, or on tape.

Solid State Detectors

It has been known for a long time that many crystals, *e.g.* diamond and cadmium sulphide exhibit electrical conductivity under the influence of radiation. Most of the materials having this property fall into the class known as semi-conductors. This pehnomenon was, however, of little general application until the advent of the transistor, and the large scale production of germanium and silicon crystals with controlled impurity content. Modern solid state detectors based on silicon and germanium are extremely useful, and are finding increased application as nuclear particle detectors—replacing gas ionisation and scintillation techniques in many cases. To understand their operation, and the terms which are used, it is necessary to understand some basic semiconductor properties.

Elements with one, two or three electrons in the outer shell are metals and good electrical conductors. Elements with a full—or nearly full—outer shell, are non-metals and are insulators. Elements which are midway between these two types, namely those with four electrons, giving a half full outer shell, are normally semiconductors *e.g.*, carbon silicon and germanium. In the normal crystal structure of these, electrons are borrowed to provide filled shells represented by the two dimension sketch (Fig. 5.8a).

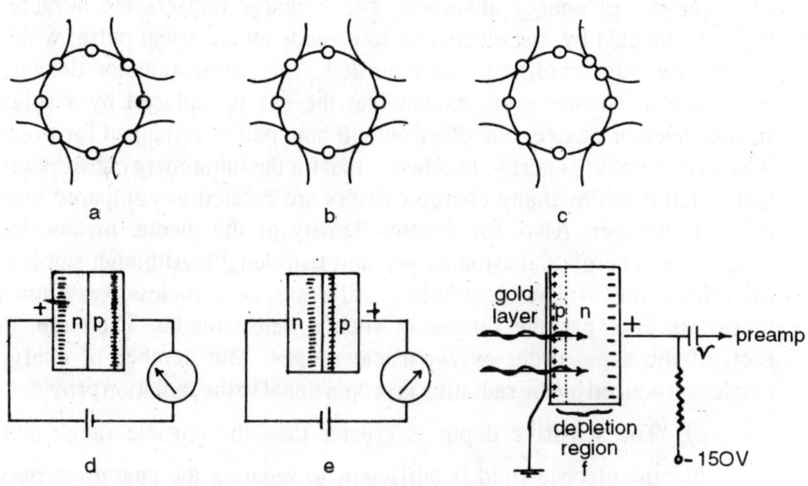

Fig. 5.8 : Principle of operation of a solid state detector.

If an impurity of similar atomic radius, with either three or five outer electrons, is introduced in trace quantities into a crystal lattice, then either an electron-deficient or electron-surplus lattice, will be produced (Fig. 5.8b and c).

An electron-surplus lattice conducts by the movement of negative electrons, and is known as an "**n-***type*" semiconductor. An electron-deficient lattice conducts by filling the "hole" with an electron from an adjacent site, thus creating a "hole" at that site, which is subsequently filled from another site. Conduction in this case is by the movement of a virtual positive "hole", and is known as a "**p**-*type*" semiconductor.

A sandwich consisting of a water of **p**-type material in contact with a water of **n**-type material constitutes a semiconductor diode. It is capable of conducting current in one direction only namely the direction of forward bias, where electrons and holes meet, and combine at the junction (Fig. 5.8c). It will not conduct in the direction of reverse bias, where the charge carriers are separated (Fig. 5.8d). This structure represents, in fact, the ordinary semiconductor diode.

To understand the operation of a semiconductor diode as a radiation detector, it is appropriate to recall the operation of the ion chamber. It will be remembered that charge carriers, in the form of ion pairs, were produced by the ionising radiation—one ion pair for every 30 eV (approx.) of energy absorbed. These charge carriers are attracted to, and collected by, the electrodes to provide an electrical pulse, which is subsequently amplified and recorded. The semiconductor detector is similar in its operation, except that the gas is replaced by a solid. In this detector device, one electron and hole pair is produced for every 3.5 electron volts of energy absorbed so that for the same energy absorption, nearly ten times as many charge carriers are created as compared with the ion chamber. Also, the greater density of the media, means that much more energy is dissipated per unit path length. Although samller, the dimensions of a semiconductor wafer are, nevertheless, very much in excess of α-particle ranges in such a dense media. They are, in fact, of the same order as β-particle ranges. The number of charge carriers produced by the radiation is proportional to the radiation provided:

a) The sensitive depth is greater than the particle range and,

b) the electric field is sufficient to separate the charge carriers before recombination can occur.

The semiconductor detector is usually constructed from an extremely thin **p**-type diffused layer on one side of an **n**-type silicon wafer (or sometimes *vice versa*), with a very thin evaporated gold electrode over the top of this thin layer. A second electrode is attached to the backing layer (Fig. 5.8f).

The detector is operated under conditions of high reverse bias, to produce a depletion region either side of the junction. The radiation eneters through the evaporated gold electrode surface. Charge carriers are formed in the depletion region; holes being swept to the negative electrode, and electrons to the positive electrode to provide a voltage pulse given by :

$$V = \frac{Q(\text{charge})}{C(\text{electrical capacity})}$$

and is of the order of millivolts per MeV dissipated.

Early detectors were suitable only for α detection, but with the development of deeper depletion zones, they are now also suitable for β and γ detection and energy spectra determination. Although operating potentials are relatively low (~ 150 volts), they produce high potential gradients of 10^3 to 10^4 V/cm in the depletion region. The pulse duration is extremely short (~10^{-9} seconds) permitting rapid counting. Energy resolution for a-particles and electrons is also extremely high (15 to 20 keV).

As production techniques improve it is highly probable that this detection device will find much more extensive usage than it does at present.

6
Nuclear Instrumentation

Introduction

It is not within the province of the chemist, for whom this book is primarily intended, to design or construct counting equipment. Consequently, the aim of this chapter is not to provided detailed circuit information, but simply to explain in general terms to the operation of the various electronic units needed to record electrical pulses generated in radiation detectors. It is hoped to provide an understanding of the functions and uses of the various units, the purpose served by the various knobs which protrude from the otherwise streamlined exteriors, to indicate the type of apparatus required for a specific purpose, and to acquaint the uninitiated with some of the terminology.

Just as a radio receiver can vary from a humble, single diode, earphone set to an elaborate multiband communications receiver, so a nuclear radiation counter can vary from the simple, fixed potential, Geiger ratemeter shown in Fig. 6.1, to a commercially manufactured multi-purpose scaler/ratemeter with some 40-50 tubes. Although the simple circuit will work, its response at higher count rates will not be linear due to variations in pulse size and its application will be very restricted. More complex instruments contain various circuit facilities, providing greater reliability and versatility to a degree proportional to their cost.

Commercially available instruments for recording the pulses produced in radiation detectors fall into two general classes:

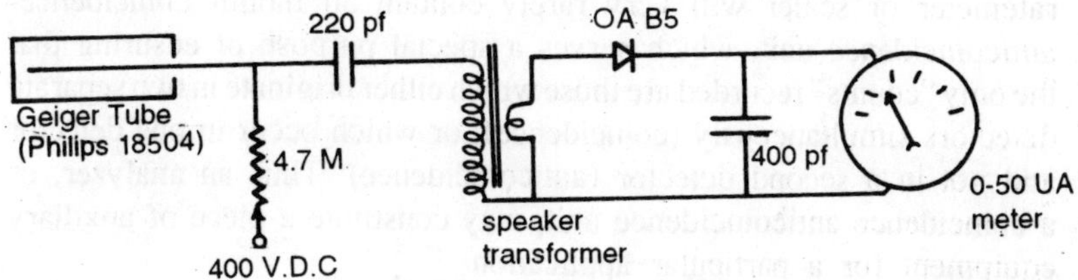

Fig. 6.1 : A simple Geiger operated ratemeter.

Scalers, which record the gross number of "counts arriving during a specific time interval, and

Ratemeters, which provide a continuous indication of the rate at which "counts" arrive.

A normal counting arrangement is summarized as a block diagram in Fig. 6.2. Besides stabilized power supplies for their own operation, the recording instruments usually contain a stabilized variable *high tension supply* (HT or EHT) for the operation of a variety of detector types. Pulses from the detector are fed into the instrument, and passed through an *amplifier* to a *discriminator,* which rejects pulses with an amplitude below a predetermined level. The pulses are then used to trigger a *square wave generator,* which produces pulses of a uniform shape and size suitable for the recording section.

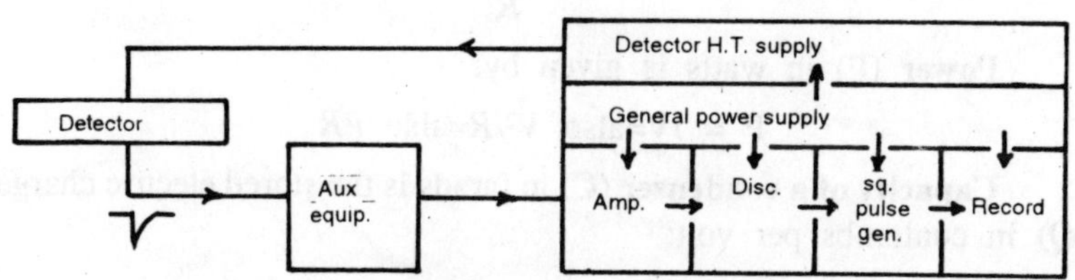

Fig. 6.2 : Block diagram of a normal counting arrangement.

In some instances, the amplification built into the instrument is not sufficient for the operation of all types of detectors, and additional external amplification is necessary as a piece of auxiliary equipment.

Infrequently, the instrument will contain an inbuilt *analyzer* to sort out and pass on only those pulses of a specific amplitude, as is necessary in the determination of radiation energy spectra. A commercial ratemeter or scaler will very rarely contain an inbuilt *coincidence-anticoincidence* unit, which serves a special purpose of ensuring that the only "counts" recorded are those which either originate in two separate detectors simultaneously (coincidence), or which occur in one detector and not in a second detector (anticoincidence). Thus an analyzer, or a coincidence anticoincidence unit, may constitute a piece of auxiliary equipment for a particular application.

The remainder of this chapter is divided into two sections:

Section A, which deals very briefly with basic electronics for those with only a scant acquaintance in this field, and

Section B, in which an outline of the principle of operation of the above mentioned units is given.

Section A. Basic Electronics

Resistors and condensors form the basic circuit links. Some of the more important equations associated with their operation are:

Ohm's law. The potential (V) in volts, developed across a resistor (R) in ohm, by a current (i) in amperes is given by:

$$V = iR$$

and the alternative form relating the current flowing through a resistor as the result of an applied potential:

$$i = \frac{V}{R}$$

Power (P) in watts is given by:

$$P = iV = \text{also } V^2/R = \text{also } i^2R$$

Capacity of a condenser (C) in farads is the stored electric charge (Q) in coulombs per volt:

$$C = \frac{Q}{V}$$

(Normal circuit capacities range several μf (10^{-6} farad) to several pf, or $\mu\mu f$ (10^{-12} farad).)

Time Constant (T) associated with a condenser-resistor circuit is the time (in seconds), which it takes the condenser (capacity C) to increase exponentially to (1 — 1/e) or approximately $^2/_3$ of the potential applied to it through a resistor (resistance R) (Fig. 6.3). It is also the time taken for the potential to drop by $^2/_3$ as a result of leakage from a charged condenser through a resistor

$$T = CR$$

Condenser-resistor circuits may be of two basic types : a potential may be applied to a condenser through a resistor (*integrating circuit*),

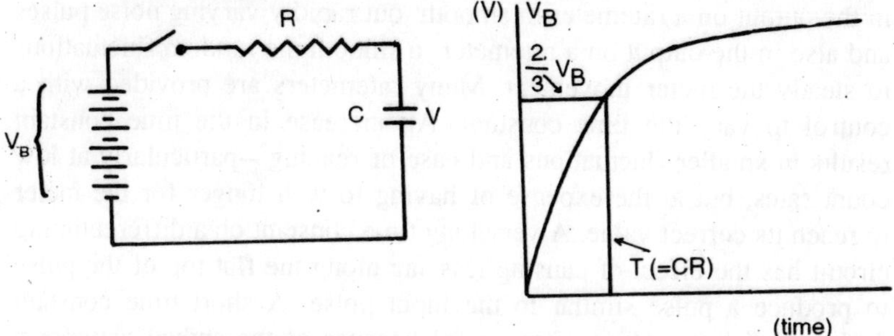

Fig. 6.3 : Time constant of a condenser-resistor circuit.

or a potential may be applied to a resistor via a condenser (*differentiating circuit*) (Fig. 6.4). The effect produced on a square shaped input pulse by each of these two circuits is also shown in Fig. 6.4. It will be noticed that the shape of the output pulse differs from that of the input—

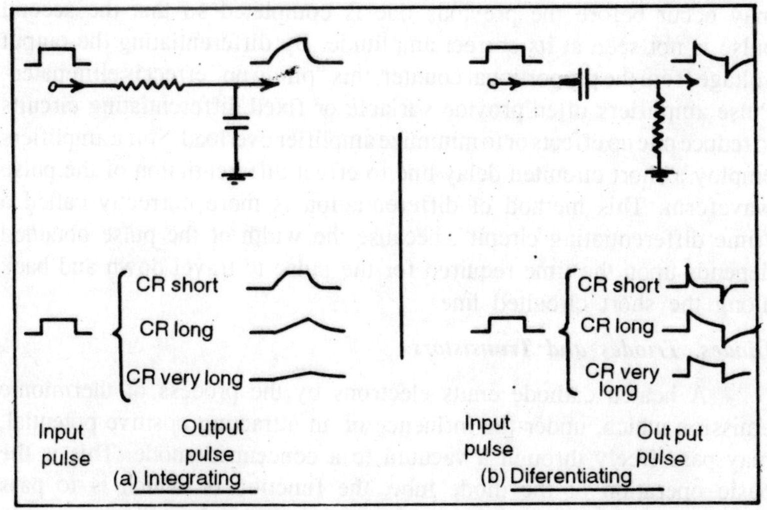

Fig. 6.4. Integrating and differentiating condenser-resistor circuit arrangements and their effect on a square wave imput pulse.

the integrating circuit mainly affecting the leading and falling edges whilst the differentiating circuit affects the flat top response, and gives rise to a negative overswing. In the case of an integrating circuit, a very long time constant ($C \times R$) results in a smoothing out effect. Such circuits are therefore used in power supply filters to remove "ripple", in an amplifier to smooth out rapidly varying noise pulses and also in the output on a ratemeter to smooth out rapidly varying noise pulses and also in the output on a ratemeter to smooth the random fluctuations to steady the meter movement. Many ratemeters are provided with a control to vary the time constant. An increase in the time constant results in smaller fluctuations and ease of reading—particularly at low count rates, but at the expense of having to wait longer for the meter to reach its correct value. A very long time constant on a differentiating circuit has the effect of causing less sag along the flat top of the pulse to produce a pulse similar to the input pulse. A short time constant differentiating circuit is often useful because of the spiked waveform obtained. For example, in Fig. 6.4b, the positive input square wave may be used to produce a spiked negative waveform suitable for triggering a univibrator at some delayed time determined by the width of the square wave. A short time constant differentiating circuit is also important for coupling a proportional counter to its pre-amplifier. Because of long collection times involved with this type of detector, a second pulse may occur before the previous one is completed so that the second pulse is not seen at its correct amplitude. By differentiating the output voltage from the proportional counter, this "piling up" effect is eliminated. Pulse amplifiers often provide variable or fixed differentiating circuits to reduce pile up effects or to minimize amplifier overload. Some amplifiers employ a short circuited delay line to effect differentiation of the pulse waveform. This method of differentiation is more correctly called a "time differentiating circuit", because the width of the pulse obtained depends upon the time required for the pulse to travel down and back along the short circuited line.

Diodes, Triodes and Transistors

A heated cathode emits electrons by the process of thermionic emission which, under the influence of an attracting positive potential, may pass freely through a vacuum to a concentric anode. This is the basic operation of the diode tube, the function of which is to pass current in one direction only. One of its principal applications is in power supplies for the production of a direct current from the alternating current output of a transformer. The solid state equivalent is the germanium or silicon diode, which consists of a junction between an **n**- and a **p**-type semiconductor; the basic operation of this has been previously described.

If a wire mesh, or "grid" a it is called, is placed between the cathode and anode of a vacuum type diode, the electron flow can be controlled by the application of a potential to this grid. This is the principle of operation of the triode used as an amplifier. Additional grids can, and are, frequently used (tetrodes and pentodes). These will not be discussed here since their operation is similar to that of the triode. The transistor equivalent of the triode is a three layer semiconductor device with alternate **p**-, **n**-, and **p**- (or **n**-, **p**-, and **n**-) type layers. The three layers are referred to as the "emitter", "base", and "collector" — corresponding to the cathode, grid, and plate in the triode. In a transistor, the emitter—collector current is a function of the emitter—base current, so that besides their lower operating potentials and obvious difference in physical construction, the fundamental difference in their mode of operation is that transistors are current sensitive devices, whilst triodes are voltage sensitive devices (Fig. 6.5 a). The two do, however, perform similar functions as amplifiers and are used in analogous circuit arrangements to perform similar operations. Although transistors find wide use, only the operational of the hot cathode tubes will be described in the text for the sake of brevity and simplicity, and in subsequent circuits only the vacuum tube counterparts will be given.

In Fig. 6.5b, the usual grid input and anode load arrangement is shown with its transistor counterpart. A positive pulse applied to the grid increases the anode current, and therefore by Ohm's law, increases the potential across the anode load resistor. The top end of this resistor is fed from a constant supply potential, and therefore the potential at the anode drops. The overall effect is an amplified pulse of reversed polarity.

Since it is the relative potential difference between the cathode and grid which effects triode operation, it is possible to apply the pulse to the cathode with the grid at constant potential (Fig. 6.5c). A positive pulse on the cathode will have the effect of making the grid relatively more negative, decreasing the anode current, and producing an amplified pulse of the same polarity.

In the *cathode follower* circuit (Fig. 6.5d) the input pulse is applied to the gird and the output taken from a cathode load resistor. Since

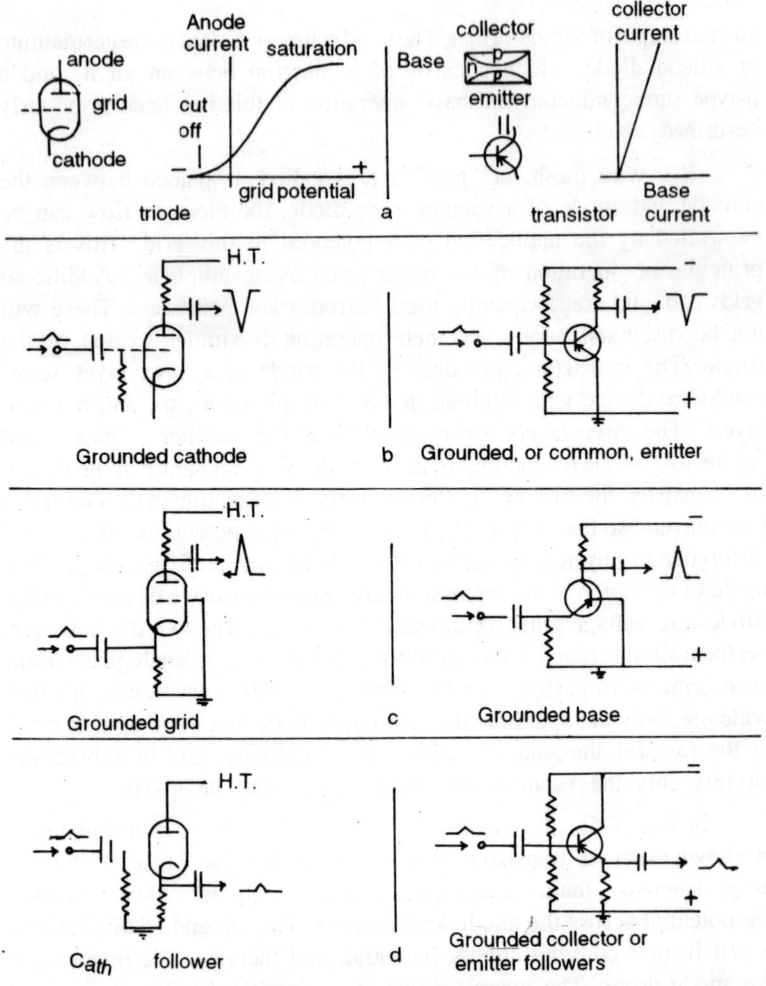

Fig. 6.5 : Comparison of the operation of vacuum tubes and transistors.

the output pulse subtracts from the cathode-grid potential difference, the voltage gain cannot exceed unity. The purpose of this circuit is not to achieve a voltage amplification, but to provide an impedance transformation from a high resistance detector to a low impedance cable—which has an inherent capacity—to the main amplifier so that the pulse may be transferred without appreciable loss or distortion due to the time constant (CR) of the system. It is quite common practice, therefore, to find commercial amplifiers split into two parts; the first, a small cathode follower unit connected as close as possible to the detector and the main amplifier situated at some more convenient position.

Electrometer tubes represent a particular variety of vacuum tubes worthy of special mention. They are essentially ordinary amplifying tubes, in which special methods of construction and reduced operating potentials reduce grid to cathode leakage current to values of less than 10^{-12} ampere. This permits the use of extremely high values of grid input resistance (10^{10} to 10^{12} ohms) which would otherwise have been shunted by the grid to cathode leakage. The use of such high value resistors allows appreciable grid potentials to be achieved from extremely small currents ($V=iR$). Electrometer tubes are commonly encountered in such instruments as ion chamber dosimeters, mass spectrometer ion current measuring circuits, and in charge sensitive amplifiers for solid state detectors.

Section B. Circuit Units

The basic principle behind the operation of some of the more common units used in nuclear instrumentation is now described. Where circuits are given as examples they will be over-simplified, in order to illustrate more clearly the principles of their operation.

Stabilised Power Supplies which are an integral part of most scalers and ratemeters consist of three sections:

(*i*) A transformer rectifier stage to produce direct current,

(*ii*) A smoothing circuit usually using a series inductance, but a resistor may also be used together with shunt capacitors—to smooth output from the rectifier, and

(*iii*) A voltage stabiliser circuit.

The latter usually consists of a "pass" tube in series with the high tension line and is controlled by an amplifier, which in turn is controlled by the pass tube output. In many respects, the pass tube acts as a variable resistor, the resistivity of which is a function of its won output (Fig. 6.6). The cathode of the amplifier is held at a fixed operating potential above ground potential by a gas discharge voltage reference tube.

The extra high tension (EHT) required for the operation of various detectors may be derived by a similar means form a high voltage power transformer, but it is frequently obtained from a high frequency (about 100 ke/s oscillator-transformer unit).

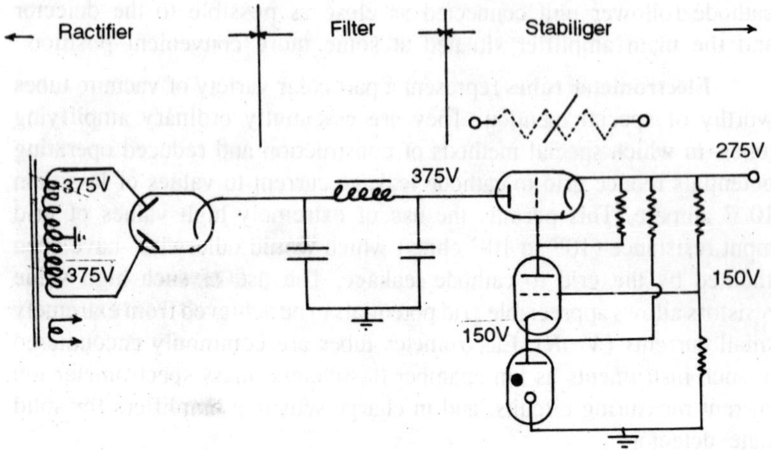

Fig. 6.6 : Simplified diagram of a regulated direct current power supply.

Amplifiers. Some degree of pulse amplification is found in most ratemeters and scalers. The amplification provided depends on the particular model, and is invariably ample for Geiger detectors. It is sometimes sufficient for scintillation detectors but infrequently sufficient for proportional, ion chamber, or solid state detectors. An additional external amplifier is then required. Amplifiers are usually resistance—capacitance coupled types, which can shape and amplify either positive or negative input pulses uniformly over a wide frequency range. For pulse spectrometry work, amplifier linearity (*i.e.* proportionality between input signal and amplified output signal) is important.

The amplification or "gain" control, has positions indicated by " × 10 or × 1000", etc. in most modern equipment. Some equipment is calibrated in terms of decibels—a heritage of audio amplifier tradition. Bels and decibels are a measure of the ratio of output and input powers.

$$\text{Number of bels} = \log_{10} \frac{\text{power output}}{\text{power input}}$$

$$= \log_{10} \left(\frac{V_0}{V_i}\right)^2$$

$$= 2 \log_{10} \frac{V_0}{V_i}$$

$$\text{Number of descibles} = 20 \log_{10} \frac{V_0}{V_i}$$

The relation between voltage gain and decibels is thus

Voltage gain	*Decibels*
2	6
10	20
100	40
1000	60

A convention practiced by a few manufacturers is to calibrate amplifiers in terms of attenuation, which is the opposite of amplification. Such an instrument will have a stated gross gain of, say, 1000 (60 dB), and the "attenuation" setting must be subtracted from this to arrive at the net gain.

Discriminator and Analyser Operation. Almost all scalers and ratemeters are equipped with a variable discriminator and, rarely, with a variable analyser.

The function of the discriminator, which is normally calibrated from 0 to 50 volts, is to discriminate against pulses below a certain selected voltage amplitude and to pass only those above this level to the counting section. It may thus be used to eliminate low amplitude electrical noise which is frequently present. It may also be used to obtain a measurement of the size of the incoming pulses by increasing the discriminator setting until they are cut off. The simplest possible discriminator is an amplifying tube to which is applied a negative grid bias in excess of the cut-off bias. Pulses which do not exceed this "over bias" value do not bring the grid potential up to its operating region and, therefore, are not passed on. Over-biassed diodes can also be used.

The discriminator is frequently incorporated as an over-biassed amplifier stage prior to a univibrator type of square wave generator (Fig. 6.7b), or directly on the "trigger" potential bias control of the Schmitt trigger square wave pulse generating circuit (Fig 6.7d).

An analyser serves the purpose of passing pulses of a certain *specific size* (amplitude). Analysers may be either single channel, which are used to "scan" a pulse spectrum, or multichannel, in which the pulse spectrum is sorted out with pulses of different amplitude being recorded in one or other of the many channels provided. A single channel pulse analyser is simply a pair of discriminators set to operate at two different values, followed by an anticoincidence unit which passes on only pulses large enough to trigger the lower level discriminator—but not large enough to trigger the upper level discriminator. The voltage difference between the two discriminator settings is referred to as the *channel width.*

Square Wave Pulse Generation

The pulses actually recorded by a scaler or ratemeter are, in fact, not the amplified pulses from the detector, but are pulses of uniform shape and size artificially generated within the instrument. The incoming amplified radiation detector pulses serve to initiate the production of these artificial pulses, so that they are equal in number to the pulses from the detector. The main reason for this somewhat apparent devious procedure, is that scaling tubes require square shaped pulses for their

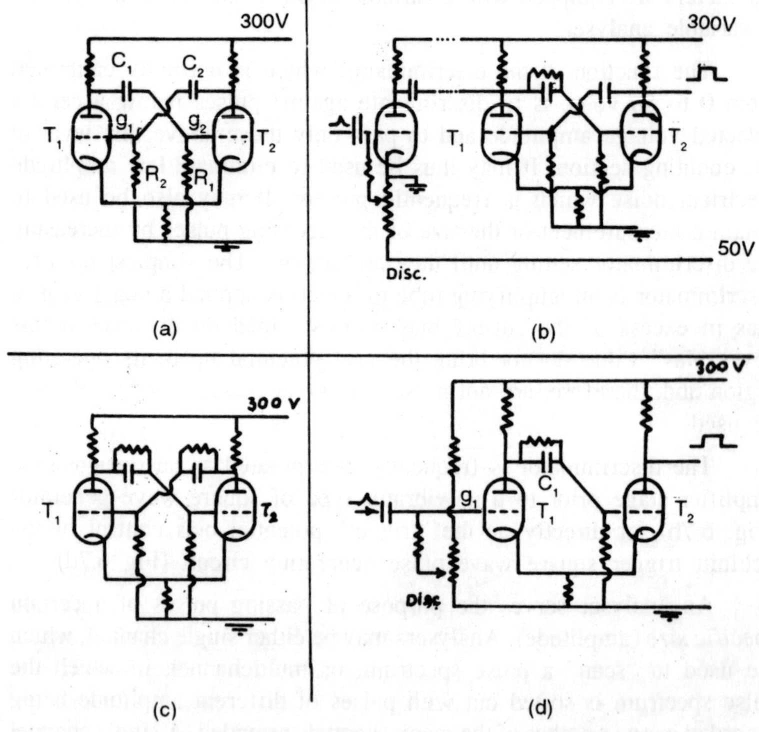

Fig. 6.7. (a) Basic circuit of the multivibrator, or square wave oscillator. (b) Operation of the mono-stable multivibrator, or univibrator. (c) The bistable multivibrator. (d) The Schmitt trigger with inbuilt discriminator.

"stepping" operation. Ratemeters which record the total charge passed, only provide a linear response to the pulse rate if pulses are of exactly the same size; *i.e.* each carrying precisely the same electric charge independent of the rate or the type of detector. Circuits providing square shaped pulses of uniform size are the univibrator and the Schmidt trigger.

Fig. 6.7a shows the basic circuit of a multivibrator, or square wave oscillator. As the potential g_1 increases, T_1 starts to conduct and an amplified negative potential appears at the anode of T_1, and at g_2. This cuts back the conduction through T_2, produces an amplified positive potential at the anode of T_2 and the gird g_1. The system is thus "regenerative", T_1 rapidly reaching saturation conduction and T_2 is totally cut off, with the potential on g_2 well below the cut-off potential. The system remains in this state, until the charge in C_1—leaking through R_1—raises the potential on g_2 above the cut-off potential. The system then regeneratively reverses itself, and so oscillates between the two states:

(a) T_1 conducting and T_2 non-conducting

(b) T_2 conducting and T_1 non-conducting.

If one of the coupling condensors, C_1, is shunted by a resistor to provide an appropriate positive potential on T_2, via a voltage divider network (Fig. 6.7b), the system will have one stable configuration—T_2 conducting—to which it will return after a single square wave generating oscillation has been initiated by a negative pulse applied to g_2 (or a positive pulse on g_1).

This is the basic circuit of the mono-stable multivibrator, or univibrator as it is also known, which is used to produce square wave pulses. If the second coupling condenser C_2 is shunted by a resistor, a bi-stable configuration is produced (Fig. 6.7c). This is not used in pulse shaping but is the basis upon which binary—or scale of two—scaling systems are based.

The basic circuit of the Schmitt trigger with inbuilt discriminator on g_1 is shown in Fig. 6.7d. In the steady state T_2 is conducting and T_1 non-conducting. An increasing potential (pulse) applied to g_1, starts T_1 conducting at a g_1 potential of about 100 volts, regeneration sets in and conduction shifts rapidly from T_2 to T_1 where it remains for a period of time determined by the value of C_1.

Coincidence—anticoincidence. A coincidence unit is a device which passes a single pulse to the recording section when two or more separate pulses are fed in simultaneously. Applications include the study of the directional properties of radiation using two or more detectors, and the study of many radioactive decay mechanisms. It operates by the application of the two pulses to separate control grids of a multigrid tube (Fig. 6.8a), or to each of the grids of two separate tubes which have a common anode load resistor (Rossi circuit) (Fig. 6.8b). In each of the above cases, the anode output pulse is fed to a discriminator, set to operate only from the larger pulses which result from the simultaneous application of pulses to both girds.

Anticoincidence circuits are used in the pulse analyser circuits mentioned earlier, and in other counting arrangements, the most important of which is the use of cosmic ray and natural background "guard" rings to reduce the background count of low background counters. The principle of operation of these counters, lies in the fact that the mini detector will record both the true β or α count from the source, and also a cosmic or terrestrial background radiation, while the guard ring will record only the background. If the output from the central detector is arranged in anticoincidence with the guard ring; *i.e.* so that the only pulses passed to the recording section are those *not* accompanied by a simultaneous pulse from the guard ring, the background count in the central detector will be considerably reduced.

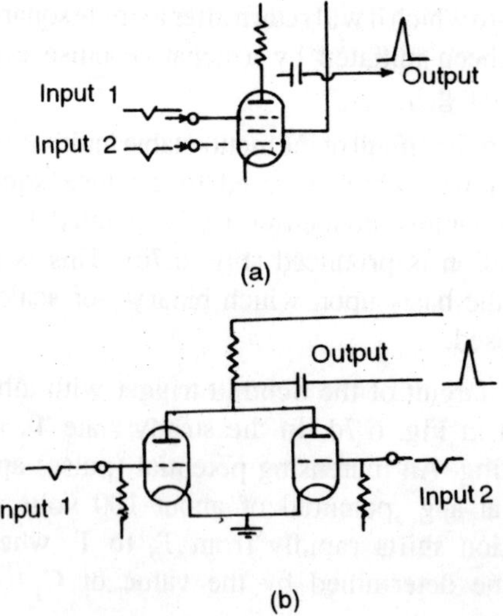

Fig. 6.8. Typical circuits used in coincidence counting applications.

The operation of the anticoincidence circuit is basically similar to that for coincidence counting, except that the polarity of one of the two pulses is reversed before being applied to one of the two control girds, thereby nullifying the effect of a pulse on the other. In practice, the pulse from the central detector must be delayed slightly in time and of shorter duration than the nullifying pulse, so that the nullifying pulse may be fully effective.

Ratemeters are units which receive pulses and—some-what like a power supply—convert them into a steady direct current, for display on a meter. This is accomplished with a rectifier and integrating circuit having a time constant which is long compared with the time interval between pulses. The time constant necessary is inversely proportional to the count rate. Provided the pulse sizes are constant, the meter deflection will be proportional to the pulse rate. The normal controls to be expected with a ratemeter section are:

(*i*) a meter range selection switch, and

(*ii*) a time constant control.

Scaling Units. The display of pulses accumulated over a period of time may be accomplished in many ways *e.g.* electromechanical registers, neon indicator tubes coupled through a binary (scale of two circuit) or a heated cathode tube wherein the electron beam is deflected progressively to different positions.

Of the various alternative types of scalers which are available the binary types have the highest counting speeds. Count rates of the order of 10^8 counts per second are obtainable with commercially available transistorized equipment.

The method which gives counting speeds suitable for general purpose work (up to 3,000 per second) coupled with reliability and low cost is the "Dekatron" tube. It consists of 30 wire cathodes arranged around a central anode (Fig. 6.9). Each third cathode corresponds to a numeral position between which are the "first guides" and "second guides". All of the first guides are internally connected and attached to a single base pin. This is also the case with the second guides. The numeral positions are also internally connected with the exception of the "O" position which is brought out via a separate connection, for the twofold purpose of passing each tenth pulse to the next dekatron, and to facilitate resetting the gas discharge to the zero position.

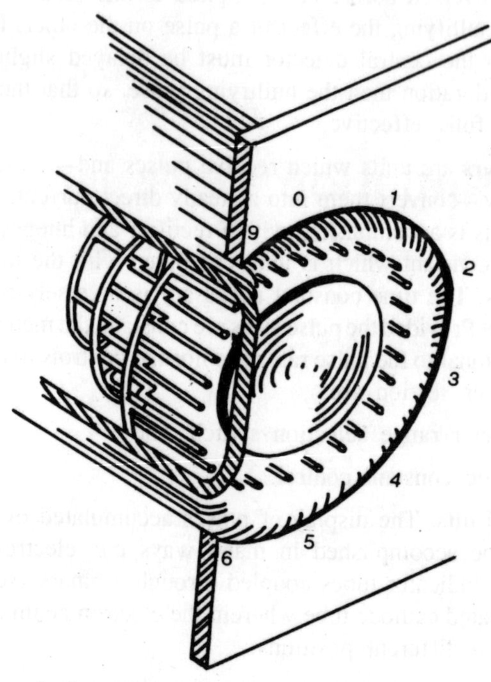

Fig. 6.9 : Construction of the "Dekatron" scaling tube.

A visible gas discharge between the anode and one of the numeral positions is moved via the guides to the next numeral position by the application of a large square shaped pulse, first to the first guides then to the second guides.

7
Radiation Chemistry

Introduction

One of the properties of nuclear radiations discussed in Chapter 4, was their ability to cause ionisation and electronic excitation in materials through which they pass. Radiation chemistry is the study of the chemical effects produced by ionising radiations of either nuclear origin, e.g. α-, β- or γ- radiation, or from artificial sources such as an X-ray machine or particle accelerators. A similar branch of chemistry is that of photochemistry, which is concerned with chemical effects resulting from absorption of lower energy photons in the visible and ultraviolet region of the spectrum. Photochemistry differs from radiation chemistry in that each visible or ultra-violet photon leads to one electron excitation, whereas in radiation chemistry a single photon or charged particle produces multiple excitations and ionisations along its path :

$$A \rightsquigarrow A^* \text{ excitation}$$

$$\text{and } A \rightsquigarrow A^+ + e^- \text{ ionisation}$$

where the symbol, $\rightsquigarrow$, indicates a radiation induced reaction, just as the symbol, $\xrightarrow{h\nu}$, is used to indicate a photochemical reaction.

Ionisation may be followed by neutralisation processes which lead to excited species :

$$A^+ + e^- \rightarrow A^{**} \rightarrow A^*$$

where by convention a double asterisk indicates a higher excited species.

In the case of molecules, dissociation of either the ionic or the excited species may occur. Examples of such processes will be given later.

An interesting point about radiation chemistry is that the outcome of an ensuing chemical reaction may depend, not only on the material irradiated, but also upon the type of radiation used. This is due to the fact that different types of radiation will produce different concentrations of activated species. α-particles, with their high energy and short range, produce a much greater concentration of ions and excited molecular species which interact with one another, than do β-particles, which have a much longer path and larger proportion of long branches, or "spurs" as they are called. These spurs are due to ionisation by secondary electrons. γ-photons which bring about ionisation via high speed electrons from an initial Compton, photo-electric, or pair-production process (Fig. 4.6) will produce a similar effect to that which might be expected from β-particles spontaneously appearing in the material (Fig. 7.1).

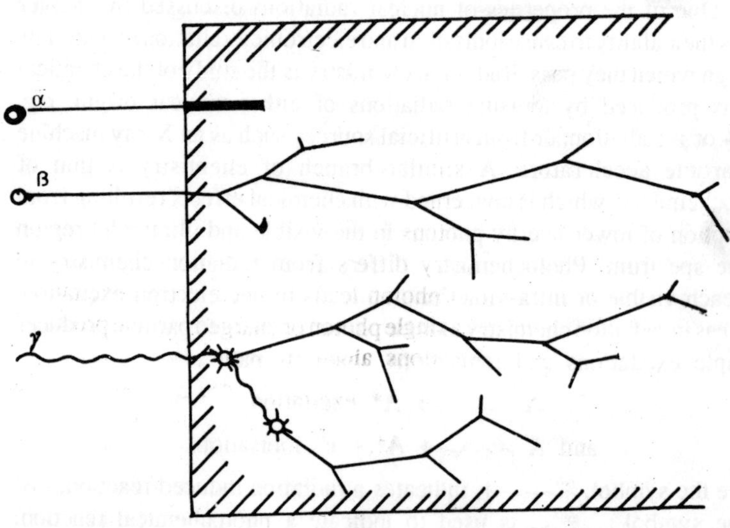

Fig. 7.1 : Pictorial representation of the interaction of α-, β-, and g-radiation with matter.

(A) Linear Energy Transfer (LET)

The amount of energy transferred to the medium is placed on a quantitative basis by a unit called *linear energy transfer* (LET) defined as the linear (*e.g.* distance) rate at which energy is lost by radiation traversing a material medium. The usual units are kiloelectron volts per micron* (keV/μ). The LET, which is, of course, a function of the type of radiation and material traversed, is also a function of the energy of the radiation, since the number of ion pairs produced per unit path length increases as the particle velocity decreases (Chapter 4, Fig. 4.2). To avoid undue complication, average LET values for a given type of radiation—with a stated initial energy in a particular medium—are usually quoted.

An α- particle has an average LET value considerably higher than a β- particle, whilst X-rays and γ-rays, which produce ionisation via secondary electrons with an average velocity somewhat less than β-particles, will have a LET value slightly higher than that of βs. Some typical LET values are given in Table 7.1.

Table 7.1

Energy and type of radiation	*LET value in water (keV/μ)*
5 MeV, α	140
2 MeV, β	0.2
1.25 MeV, g (^{60}Co)	0.3
200 KeV, X-rays	3

G value

Where a single atomic species (*e.g.* argon) is involved in absorbing radiation, recombination of an ion and electron occurs and there is no residual effect. Where molecules are involved, however, chemical bonds are frequently raptured and there is a residual chemical effect. A term to describe this in quantitative terms, is the G *value*, which denotes the number of molecules changed for each 100 electron volts of energy absorbed. In many radiation chemical reactions, a large number of different molecules, radicals and ions may be involved, and a subscript is frequently employed to denote the species to which referencc is made *e.g.* G_{H_2O}, G_{OH}, or $G_{Fe^{3+}}$.

It should be remembered that the G value will be a function of the ionisation density produced by the radiation. For example, the G value for the production of Fe^{3+} ($G_{Fe^{3+}}$) in the Fricke (ferrous sulphate) dosimeter varies from 5.1 for 5.3 MeV α-particles, to 15.5 for 1.25 MeV (^{60}Co) γ-rays. The reason for this, together with the chemical reactions which bring about the oxidation of the ferrous ion, are described later.

* 1 micron (μ) = 10^{-4} cm.

Before looking in more detail at the chemical effects produced by the absorption of radiation, it is appropriate to introduce some units which define the amount of radiation (the roentgen) and the amount of radiation absorbed by a medium (the rad).

Earlier, in Chapter 1, the unit of γ- or X-ray radiation, the *roentgen* (r) was defined as *that quantity of X-rays or γ-rays which will produce, as a consequence of ionisation, one e.s.u. of electric charge of either sign, in one cc of dry air measured at S.T.P.* (*i.e.* in 0.001293 g of dry air). Although the definition restricts the unit to electromagnetic radiation, it is also frequently applied in practice to β-radiation, and β-γ ion chamber dosimeters are frequently used for the purpose of monitoring laboratory radiation levels.

Some prospective on the magnitude of the roentgen as a unit can be gauged from the fact that exposure of the whole human body to a total of 600 r is fatal, while the normal annual whole body irradiation from natural sources, mainly our own potasium-40 and carbon-14 content, together with cosmic and terrestrial background is equivalent to 0.2 r of whole body irradiation, and that the accepted limit for occupational exposure is 0.1 r per week.

Besides this unit which defines the quantity of radiation, it is also necessary to have a unit to describe the *absorbed* radiation dose. This is the *rad* and is defined as *the absorbed dose of* **any** *ionising radiation which is accompanied by the liberation of* 100 *ergs of energy per gram of absorbing material* (=$6.24 . 10^{13}$ e V/g in electron volt energy units). It will be noticed that this unit is applied to all types of radiation. Furthermore, it is applied not only for radiation sources external to the media absorbing the energy, but also where a radioactive isotope is contained within the material, *e.g.* a solution containing an α emitting nuclide. In such cases where all of the particle energy is absorbed, the dose is readily calculated from a knowledge of the specific activity, *i.e.* the number of particles per unit time and their energy (remembering that 1 MeV = $1.602 \cdot 10^{-6}$ erg).

Difficulty is frequently encountered in understanding the different functions served by the roentgen and the rad. These are best understood by considering a hypothetical example, where a radiation source and ion chamber are placed on opposite sides of an evacuated flask. If the ion chamber records a radiation intensity of 1 roentgen per hour, then 1 roentgen of radiation is passing through the vacuum each hour— but no radiation is being absorbed by the void, *i.e.* the absorbed dose is zero.

Another way to appreciate the inter-relationship between the units is to determine the absorbed dose in air exposed to 1 r of radiation. Assuming an electronic charge of $4.80 \cdot 10^{-10}$ e.s.u. and that an average value of 34 eV of energy is necessary to produce each ion pair, then 1 e.s.u. of charge (which corresponds to the production of $1/4.40 . 10^{-10}$ ion pairs) requires a total energy absorption of

$$\frac{34}{4.80 \times 10^{-10}} \text{eV} = 7.08 \times 10^{10} \text{eV}$$

$$= 11.34 \times 10^{-2} \text{erg}$$

$$(1 \text{ eV} = 1.602 \times 10^{-12} \text{erg}).$$

This is in 0.00129 g of air. The energy absorbed in 1 g would be 88 erg, *i.e.* the absorbed dose in *air* exposed to 1 r is 0.88 rad.

In the case of water, the value will, of course, be different. In fact, 1 r of γ-radiation will result in approximately 97 ergs being absorbed in 1 g of water, *i.e.* the absorbed dose in water exposed to 1 r is 0.97 rad.

Radiation Dosimetry

The measurement of radiation dose —either absorbed dose or exposure dose—can be made by a variety of methods. Some of the more important ones are outlined below.

(A) Ion Chamber Measurements

The operation of an ion chamber as a measuring instrument has been given in Chapter 5. As the roentgen is defined in terms of the ionisation produced in air, the direct measurement of this quantity using an ion chamber is an obvious approach. Unfortunately the ionisation currents, which are small, impose some practical limitations. For example, 1 r per hour implies the release of 1 e.s.u. of electric charge in each cc each hour. In an ion chamber with a volume of 100 cc this would amount to 100 e.s.u. per hour, or 0.028 e.s.u. of charge per second, which is equivalent to approximately $9 \cdot 10^{-12}$ coulomb per second or $9 \cdot 10^{-12}$ ampere.

Specially constructed chambers which permit a collimated beam of radiation to produce ionisation within a specified volume of the chamber without scattering, and without the production of secondary electrons from electrodes or walls, can be sued for the absolute determination of radiation levels. In practice, however, it is very much more common to use much less elaborate equipment previously calibrated against a standard source of radiation. Ion chambers respond to a very range of radiation levels and are particularly useful as laboratory monitors.

(B) Calorimetry

Because the rad is defined in terms of absorbed energy, the measurement of heat produced by the absorption of radiation in a graphite or metallic block is obviously the direct approach to the absolute measurement of absorbed dose, but because the heat production is so low (1 rad = 100 ergs/g = $2.39 \cdot 10^{-6}$ cal/g) it is only practicable at high levels of radiation intensity.

(C) Chemical Dosimetry

A very large number of chemical dosimeters which utilize a radiation induced chemical reaction have been evolved. Whilst they do not provide an absolute standard for measurement, they are very convenient secondary standards. In general, they employ conventional chemical techniques and are suitable for medium and high dose regions. Special analytical techniques have been evolved in some cases to extend their usefulness into the lower radiation regions.

Probably the best known of the chemical dosimeters is the Fricke (ferrous sulphate) dosimeter which is based on the radiation induced oxidation of ferrous ion to ferric ion. An air saturated solution 0.001 *M* with respect to ferrous ammonium sulphate, 0.001 *M* with respect to sodium chloride and 0.4 *M* (0.8 *N*) with respect to sulphuric acid is used. For every 100 eV of γ energy absorbed, 15.5 ferric ions are produced (*i.e.* $G_{Fe^{3+}}$ = 15.5.). The lower limit of radiation which can be monitored thus depends upon the chemical detection methods available. The lower limit when spectrophotometric means are employed is several hundred rads, but if radioactive iron -59 is used as a tracer, subsequently isolated as the ferric state, and the more sensitive radiochemical assay techniques are used, the lower limit can be reduced to only a few rads.

Very many other inorganic and organic reaction can also be employed. One which required special mention, because of its particular application, is the darkening of a photographic film. This is the basis of "film badges" worn by persons working in radiation areas to monitor low exposure doses accumulated over a period of time, and which serves a similar purpose to the ion chamber pocket dosimeters mentioned in Chapter 5 .

Chemical Effects Resulting from the Absorption of Radiation

A great number of chemical reactions result from the absorption of high energy radiations and a large amount of information concerning the outcome of many individual radiation induced reaction is recorded in the chemical literature. In some cases, detailed mechanisms have been elucidated, but in a great many, only the net effect has been established and considerable doubt exists concerning the mechanism. Radiation chemical reactions are not straight-forward reactions as are many in organic and inorganic chemistry, but are rather to be interpreted in the light of certain guiding principles. It is neither possible, nor of great value, to enumerate the various reactions which have been observed, in a book of this nature. What can be done is to indicate the more important mechanisms whereby radiation induced reactions are interpreted, and to provide examples of a few selected typical reaction to illustrate these principles.

Radiation chemistry is, in effect, the chemistry of molecular ions, excited molecules, and radicals which are produced by the passage of ionising radiation through a medium.

(A) Primary Mechanisms in Molecules

As previously stated, the initial effect of radiation is the production of both *molecular ions*—or *ion molecules* as they are frequently termed—and electronically *excited molecules*.

$$A \rightsquigarrow A^+ + e^- \quad \text{(ionisation)}$$
$$A \rightsquigarrow A^* \quad \text{(excitation)}$$

(B) Secondary Mechanisms in Ionised and Excited Molecules

(a) Ions:

An ion may be neutralised to produce an electronically excited molecule:

$$A^+ + e^- \rightarrow A^{**} \rightarrow A^*$$

where A** indicates an upper excited molecular state, and A* indicates a lower excited molecular state.

The charge may also be transferred to another molecule with a lower ionisation potential:

$$A^+ + B \rightarrow A + B^+$$

Molecular ions may enter directly into a chemical reaction, or, if they are in an initially excited state, they may first dissociate into a radical and a radical ion which subsequently enter into chemical reactions

$$(A^+)^* \rightarrow B\cdot + C^{+\cdot}$$

(b) Excited Molecules

Excitation energy can simply be emitted as a photon (*hv*); the molecule thereby reaching its ground state *i.e.* fluorescence.

$$A^* \rightarrow A_t + hv$$

Alternatively, the excitation energy may be transferred to another molecule which has lower energy levels

$$A^* + B \rightarrow B^* + A$$

Excited molecules may also enter directly into chemical reactions, or via *free radicals* produced by the rupture of a chemical bond.

$$A^* \rightarrow B\cdot + C\cdot$$

The condition necessary for bond rupture is that the available excitation energy is at least that of the bond energy. The pair of electrons which form the bond are divided between the radicals, are indicated by "dots" and remain as potential bond linkages. An important feature of the process is that it is not necessarily the bond attaching the atom which received the energy that breaks. Energy is distributed throughout the molecule with a greater probability of bond fracture the lower the bond energy. Some of the more common bond energies are shown in Fig. 7.2.

(C) Ensuing Chemical Reactions

The important chemically reacting species which are produced by both primary and secondary effects (Fig. 7.3) that are considered in subsequent reactions are:

(a) Ion molecules,

(b) Excited molecules, and

(c) Free radicals.

(a) Ion Molecules

Besides the charge neutralisation and charge transfer processes mentioned above, activated molecular ions may also dissociate into either a radical ion and a radical (also mentioned above) which subsequently react further, or into molecular ion and another molecule.

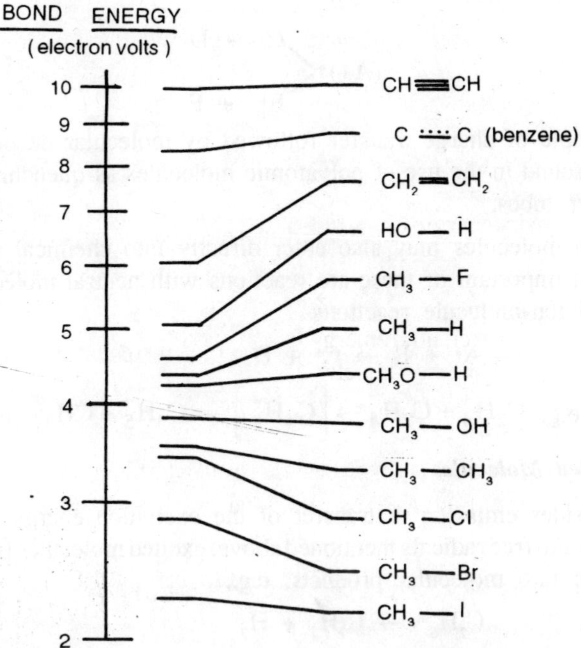

Fig. 7.2. Values of bond energies for some common systems.

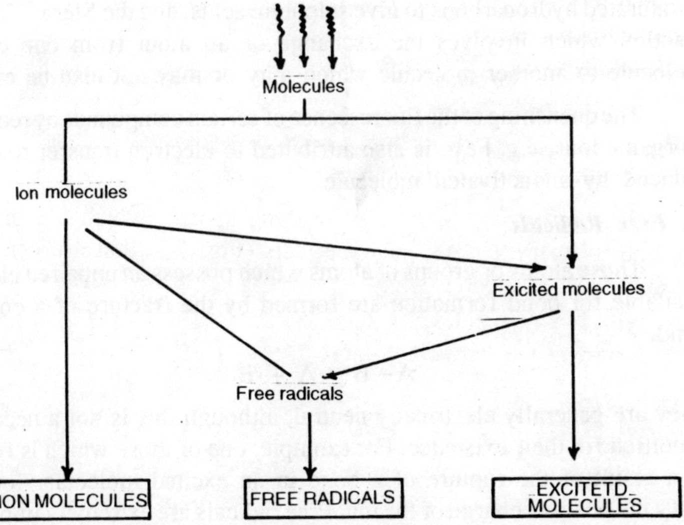

Fig. 7.3 : The production of Ion Molecules, Free Radicals and Excited Molecules by high energy radiation.

$$(A+)^* \begin{cases} C^{+\cdot} + D\cdot \\ E^{+} + F \end{cases}$$

An example of charge transfer followed by molecular decomposition is to be found in the use of polyatomic molecules as quenching agents in Geiger tubes.

Ion molecules may also enter directly into chemical reactions. The most important of these are reactions with neutral molecules, the so-called ion-molecule reactions

$$A^+ + B \rightarrow C^+ + D$$

$$\text{e.g. } C_2H_4^{\dot{+}} + C_2H_4 \rightarrow \left[C_4H_8^{\dot{+}}\right] \rightarrow C_3H_5^{\dot{+}} + CH_3\cdot$$

(b) Excited Molecules

Besides emission or transfer of the excitation energy, and dissociation into free radicals mentioned above, excited molecules frequently dissociate into molecular products, *e.g.*,

$$C_2H_6^* \rightarrow C_2H_4 + H_2$$

Examples of other less common types of excited molecule reactions include addition reactions *e.g.* the addition of activated sulphur dioxide to saturated hydrocarbons to give sulphinic acids, and the Stern—Volmer reaction which involves the exchange of an atom from one excited molecule to another molecule which may or may not also be excited.

The quenching of the fluorescence of certain compounds by reducible inorganic ions, *e.g.* Fe^{3+}, is also attributed to electron transfer reactions induced by an activated molecule.

(c) Free Radicals

Those atoms or groups of atoms which possess an unpaired electron available for bond formation are formed by the fracture of a covalent bond.

$$A-B \rightleftarrows A\cdot + \cdot B$$

They are generally electrically neutral, although this is not a necessary condition for their existence. For example, one of those which is formed as a result of the rupture of a bond in an excited molecular ion will carry the positive charge of the ion. Free radicals are extremely important in the interpretation of radiation chemistry, just as they are important in the allied field of photochemistry. Their stability varies over extremely wide limits: H· and ·OH, the basis of aqueous systems, are extremely reactive. Reactivity among the organic radicals decreases with increasing size, the $\cdot CH_3$ radical being the most reactive and capable of only transient existence while at the other extreme. $Ph_3C\cdot$ is stable. With the halogens, reactivity decreases with decreasing electronegativity. The more reactive the radical, the less discerning it is in its attack on a neutral molecule. Free radical chemistry is a branch of chemistry on its own. The reactions are independent of the way in which the radicals were formed and all that will be attempted here is to give a few brief examples of free radical mechanisms which are pertinent to some typical radiation chemical systems:

(1) Combine together and

(a) produce a neutral molecule

e.g. $\cdot CH_3 + \cdot CH_3 \rightarrow C_2H_6$

$H\cdot + \cdot OH \rightarrow H_2O$

or

(b) combine and disproportionate

e.g. $\cdot C_2H_5 + \cdot C_2H_5 \rightarrow C_2H_6 + C_2H_4$

(2) React with a neutral molecule to

(a) produce a different radical

$H\cdot + CH_3I \rightarrow \cdot CH_3 + HI,$

or

(b) produce a similar type of radical resulting in a chemical chain reaction; for example the polymerization of styrene

$$-CH_2-\underset{\displaystyle Ph}{\overset{|}{CH}}\cdot + CH_2{=}\underset{\displaystyle Ph}{\overset{|}{CH}} \rightarrow -CH_2-\underset{\displaystyle Ph}{\overset{|}{CH}}-CH_2-\underset{\displaystyle Ph}{\overset{|}{CH}}\cdot \text{ etc.}$$

(3) Promote an electron transfer process.

e.g. $\cdot OH + Fe^{2+} \rightarrow Fe^{3+} + OH^-$

Radical scavengers. Radiation induced reactions are frequently difficult to interpret because of the large number of species which may be present. Certain materials can be used to scavenge free radicals by preferential reaction with them, thus inhibiting that part of the reaction attributable to free radicals. A very commonly used radical scavenger is iodine. Below 10^{-3} molar it has very little effect on the outcome of a reaction which takes place exclusively in the radiation tracks, but has considerable effect if the reaction takes place in the bulk of the medium after diffusion of radicals. At higher iodine concentrations radical reactions taking place in the tracks prior to diffusion are affected.

$$R\cdot + I_2 \rightarrow RI + I\cdot$$

Iodine radicals are relatively stable and unreactive and they ultimately recombine with one another. Analysis of the organic iodides produced in the scavenging reaction can be used as a means of radical identification. Frequently, iodine labelled with radioactive iodine-131 is used to facilitate the final analysis.

(D) Radiation Chemical Reactions in Gases

The low density of gases as a medium results in lower LET values, a more homogeneous distribution of radiation products, and less differentiation between the effects of α- and γ-radiation. Studies in gases and vapours are of special significance because they permit a complementary and independent approach to radiation reactions by using a mass spectrometer. In this instrument, which provides a continuous sampling of the charged species present, the ions are produced by controlled energy electron bombardment. Much of the information which has been elucidated for the radiation decomposition of water, for example, has been inferred from mass spectrometric measurements on water vapour.

Of possible potential commercial importance is the polymerization of ethylene to produce polyethylene. The precise course of the reaction is not yet understood though in outline it can be represented as :

$$R\cdot + C_2H_2 = C_2H_2 \rightarrow R-CH_2-CH_2\cdot + C_2H_2 = C_2H_2 \rightarrow \text{etc.} \rightarrow \text{polymer.}$$

At atmospheric pressure and room temperature, the G value is only 11, but rises to a value of 10^4 at 240 °C and 21 atmospheres pressure.

Another gaseous system which has been the object of some investigation for potential commercial application, is the fixation of atmospheric nitrogen in the manufacture of fertilizers. The irradiation of dry nitrogen-oxygen mixture produces a mixture of nitrogen oxides, but in the presence of moisture, nitric acid is the chief product.

(E) Radiation Chemical Reaction in Aqueous Systems

By far the most important single system is that of aqueous solutions; *i.e.* the effects produced in water and substances dissolved in water by the absorption of radiation. In general, these are dilute solutions, and most of the radiation will be absorbed by the water molecules rather than by the material dissolved in the water. The subsequent chemical effects produced in the solute result form reactions with the active species produces in the water. Because of its importance, aqueous systems will be considered in some detail; first, the effect of radiation on pure (oxygen free) water will be considered, then water containing dissolved oxygen, and finally water containing other dissolved material.

The effect produced by the absorption of radiation on pure oxygen free water, is to produce H· + ·OH radicals.

$$H_2O \rightsquigarrow H\cdot + \cdot OH$$

The reaction involving electronically excited molecules to produce these radicals appears as straightforward decomposition of the excited molecule, but although there have been extensive studies, there still remains some doubt concerning the exact mechanisms for the production of these radicals from the ionic species produced by the radiation. Mass spectrometric measurements of the vapour phase suggest initial reactions along the following lines:

$$H_2O \rightsquigarrow H_2O^+ + e^-$$
$$H_2O^+ + H_2O \rightarrow H_3O^+ + \cdot OH$$
$$H_2O \rightsquigarrow OH^+ + H\cdot + e^-$$
$$H_2O \rightsquigarrow H^+ + \cdot OH + e^-$$

Mass spectrometric abundance measurements of the positive ions H_2O^+, H_3O^+, OH^+ and H^+, produced in the vapour phase are in the ratio 10:2:2:2. The reaction shown in the second equation in inferred from the fact that the H_3O^+ concentration is proportional to the square of the vapour pressure, thereby indicating a bimolecular reaction. Subsequent neutralisation of the ionic species will lead to additional radicals by the following reactions:

$$H_2O^+ + e^- \rightarrow H\cdot + \cdot OH$$
$$H_3O^+ + e^- \rightarrow 2\ H\cdot + \cdot OH$$
$$H^+ + e^- \rightarrow H_2O \rightarrow 2H\cdot + \cdot OH$$
$$OH^+ + e^- + H_20 \rightarrow H\cdot + 2\ \cdot OH$$

Ionic reactions in the liquid phase will differ somewhat from those in the vapour, although the differences in the liquid phase will be more of degree rather than of nature. The major differences to be expected in the liquid phase are:

(a). Hydration of the ions resulting in an increase in the stability of these species,

(b) Hydration of the electrons. It is now generally accepted that, in the liquid phase, electrons produced in the ionisation process are hydrated. They are thought of as an electronic charge distributed over several water molecules. They are believed to be an important reducing species, and to give rise to hydrogen radicals by the reaction

$$e^-(aq.) + H_2O \rightarrow H\cdot + OH^-$$

(c) The LET values will be very much higher with correspondingly higher concentrations of reacting species in localised areas, with a greater probability of reaction with one another.

For example, the electron will not have moved far and neutralisation of the ion would be more probable:

$$H_2O+ + e^- \rightarrow H_2O^{**} \rightarrow H\cdot + \cdot OH$$

On the other hand, a higher local concentration of H· and ·OH radicals produced by all the processes would lead to a greater probability of recombination

$$H\cdot + \cdot OH \rightarrow H_2O$$

This latter reaction is the reason why the measured G value for the decomposition of liquid water is only one-third that of water vapour. The probability of the reactions:

$$2\,H\cdot \rightarrow H_2\ .$$

and

$$2\cdot OH \rightarrow H_2O_2$$

is also increased with higher LET values. This is the reason attributed to $G_{H\cdot}$ and $G_{\cdot OH}$ values for α-radiation being lower than for β-radiation while the G_{H_2} and $G_{H_2O_2}$ values are at the same time increased.

The foregoing discussion has been included for the twofold purpose of illustrating the type of reasoning adopted to interpret reactions, and to indicate that even in such well studied—and seemingly simple—systems such as water, doubt still exists concerning the intermediates involved in the reactions.

From a practical point of view, the overall reactions leading to the production of radicals in water can be represented in a simplified form:

$$H_2O \rightsquigarrow H\cdot + \cdot OH$$

With subsequent reaction between the H· and ·OH radicals to produce H_2, H_2O_2, and H_2O:

$$2\ \cdot H \rightarrow H_2,$$
$$2\ \cdot OH \rightarrow H_2O_2 \text{ and}$$
$$H\cdot + \cdot OH \rightarrow H_2O$$

The overall reaction representing all molecular and radical species present is then:

$$x\ H_2O \rightsquigarrow aH_2 + bH_2O_2 + cH\cdot + d\cdot OH$$

Oxygenated Water

The above discussion was restricted to pure water. In water which contains dissolved oxygen, *i.e.* water that has been exposed to the air, there will be an additional effect brought about by the scavenging of H· by the oxygen present to give a perhydroxyl radical.

$$H\cdot + O_2 \rightarrow HO_2\cdot$$

The $HO_2\cdot$ radical is not as reactive as is H·, but it is a strong oxidizing radical and reacts with ·OH radicals to given oxygen, and with other $HO_2\cdot$ radicals to give oxygen and hydrogen peroxide.

$$HO_2\cdot + \cdot OH \rightarrow O_2 + H_2O$$
$$2HO_2\cdot \rightarrow O_2 + H_2O_2$$

The ultimate yield of H_2O_2 in oxygenated water is also enhanced by the fact the scavenging of H· inhibits the decomposition of H_2O_2 which would otherwise result from the reaction:

$$H\cdot + H_2O_2 \rightarrow \cdot OH + H_2O$$

Aqueous Solutions

Various compounds and ions react with one or more of the radiation products produced in water. For example, added hydrogen peroxide reduces the H· concentration by the reaction:

$$H\cdot + H_2O_2 \rightarrow \cdot OH + H_2O$$

while hydrogen, carbon monoxide and halide ions will react with ·OH:

$$\cdot OH + H_2 \rightarrow \cdot H + H_2O$$

$$\cdot OH + CO \rightarrow \cdot H + CO_2$$

$$\cdot OH + Cl^- \rightarrow Cl\cdot + OH^-$$

This latter reaction is assisted in acid media

$$\cdot OH + Cl^- + H^+ \rightarrow Cl\cdot + H_2O$$

Ferrous Sulphate Solution (Fricke Dosimeter)

The oxidation of ferrous ion to ferric ion in aerated acid media is a well established and reliable method for the measurement of radiation doses . Oxidation of the ferrous ions by the various reactive species produced by the radiation in the aqueous media proceeds in a manner such that each ·OH radical is responsible for the oxidation of one ferrous ion, each H_2O_2 molecule oxidises two ferrous ions, and each H· radical—after interaction with dissolved oxygen to produce HO_2—oxidises three ferrous ions:

$$\cdot OH \xrightarrow{+Fe^{2+}} Fe^{3+} + OH^-$$

$$H_2O_2 \xrightarrow{+Fe^{2+}} Fe^{3+} + OH + OH^-$$

$$\downarrow + Fe^{2+}$$

$$Fe^{3+} + OH^-$$

$$HO_2 \xrightarrow{+Fe^{2+}} Fe^{3+} + HO_2^-$$

$$\downarrow + H^+$$

$$H_2O_2 \xrightarrow{+Fe^{2+}} Fe^{3+} + \cdot OH + OH^-$$

$$\downarrow + Fe^{2+}$$

$$Fe^{3+} + OH^-$$

The yield of ferric ion ($G_{Fe^{3+}}$) is related to the molecular and radical yields by:

$$G_{Fe^{3+}} = G_{\cdot OH} + 2G_{H_2O_2} + 3G_H$$

The value of $G_{Fe^{3+}}$+ for γs in aerated water is 15.5 but for αs is only one-third this value due to the higher LET values.

In the absence of air, $HO_2\cdot$ is not formed and one H· radical leads to the oxidation of a single ferrous ion:

$$H\cdot + H+ + Fe^{2}+ \rightarrow Fe^{3}+ + H_2$$

The yield of ferric ion GFe³+ which has an experimentally determined value of 8.2, is then related to the molecular and radical yield by

$$G_{Fe^{3+}} = G_{\cdot OH} + 2G_{H_2O_2} + G_{H\cdot}$$

The set of chemical equations above represents the ideal situation. The presence of organic impurities can lead to a series of reactions with the ·OH radical which, in most cases, bring about the ultimate oxidation of three ferrous ions instead of one, and in some cases, *e.g.* traces of ethanol or formic acid, the *G* value for the oxidation of ferrous ion rises to values of up to 75 and 250 respectively. To avoid uncertainties due to traces of organic impurity sodium chloride is incorporated in the Fricke ferrous sulphate dosimeter. Each ·OH radical is scavenged by the chloride present to produce a chlorine radical in its place. The chlorine radical is still capable to reducing one ferrous ion, but does not enter into complex reaction with the organic impurities

$$\cdot Cl + Fe^{2+} \rightarrow Fe^{3+} + Cl^{-}$$

Other forms of aqueous oxidation—reduction dosimeters e.g. the ceric sulphate dosimeter, are also used. In this system ceric ions are reduced by $HO_2\cdot$ and H_2O_2 and cerous ions oxidized by ·OH:

$$H\cdot O_2 \longrightarrow HO_2$$

$$HO_2 \xrightarrow{+Ce^{4+}} Ce^{3+} + H^{+} + O_2$$

$$H_2O_2 \xrightarrow{+Ce^{4+}} Ce^{3+} + HO_2 + H^{+}$$

$$\downarrow {+Ce^{4+}}$$

$$Ce^{3+} + H^{+} + O_2$$

$$OH \xrightarrow{+Ce^{3+}} Ce^{4+} + H^{-}$$

The net yield of cerous ion ($G_{Ce^{3+}}$) is found by experiment to be 2.4 and is related to the molecular and radical yield by:

$$G_{Ce^{3+}} + = G_{H\cdot} + 2G_{H_2O_2} - G_{\cdot OH}$$

From a knowledge of the *G* values for ferrous oxidation with and without air, together with the value for the reduction of cerous ions, and assuming that the primary products from the irradiation of water are the three reacting species H·, ·OH, H_2O_2 plus the non reacting species H_2, it is possible to calculate from a set of simultaneous equations the individual radical and molecular yields for γ irradiation of water under Fricke dosimeter solution conditions. These are shown in Table 7.2, together with their contribution to the overall G values for the various dosimeters mentioned above.

Aqueous Solutions of Organic Compounds

The free radicals produced in water by radiation react with a number of organic compounds to produce organic radicals which subsequently undergo a variety of reactions. Succinic acid is formed from acetic acid solutions by dimerisation of $\cdot CH_2COOH$ radicals, ethanol in the

Table 7.2

G	*Numerical value*	*Contribution to G value of dosimeter*		
		Fe^{3+} *aerated*	Fe^{3+} *non-aerated*	Ce^{3+} *aerated*
G–H_2O	4.5			
G_{H_2}	0.4			
$G_{H_2O_2}$	0.8	× 2 = 1.6	× 2 = 1.6	× 2 = 1.6
$G_{H\cdot}$	3.7	× 3 = 11.1	× 1 = 3.7	× 1 = 3.7
$G_{\cdot OH}$	2.9	× 1 = 2.9	× 1 = 2.9	× 1 = (–)2.9
		$G_{Fe^{3+}}$ = 15.6	$G_{Fe^{3+}}$ = 8.2	$G_{Ce^{4+}}$ = 2.4

absence of oxygen forms CH_3CHOH radicals which subsequently react to give acetaldehyde and butane-2, 3-diol, carbohydrates undergo reactions involving oxidation, degradation and polymerisation. Chloroform, which is slightly soluble in water, is an interesting and useful example. It gives HCl, H_2O_2 and CO_2. Although the mechanism is uncertain, the high *G* value of 26.6 for HCl production in aerated solution (6.3 in deaerated solution) together with easy chemical estimation methods for HCl measurement make it very suitable as a dosimeter.

(F) Radiation Chemistry of Pure Organic Compounds

In most cases a number of products are produced from a single starting compound.

(a) Aliphatic Compounds

In aliphatic compounds, C—C and C—H bonds rapture with approximately equal probability, although tertiary C—C bonds generally break more readily than secondary C—C bonds, which in turn are more susceptible than are primary C—C bonds. The radiation products include hydrogen and hydrocarbons of both lower and higher molecular weight than the starting material, with the dimer as a frequent product. Many aliphatic systems have been studied using iodine as a radical scavenger to identify the products.

With the exception of fluorides—which possess a strong carbon—halogen bond—aliphatic halides usually fracture at the carbon—halogen bond,

$$CH_3I \rightsquigarrow \cdot CH_3 + \cdot I$$

This reaction is then followed by the various possible reactions of the free radicals:

$$\cdot CH_3 + \cdot I \rightarrow CH_3I \text{ (recombination)}$$

$$2\ \cdot CH_3 \rightarrow C_2H_6$$

$$2 \cdot I \rightarrow I_2$$

while in the case of carbon tetrachloride:

$$CCl_4 \rightsquigarrow \cdot CCl_3 + \cdot Cl$$

followed by:

$$\cdot CCl_3 + \cdot Cl \rightarrow CCl_4 \text{ (recombination)}$$

$$2 \cdot CCl_3 \rightarrow C_2Cl_6$$

$$2\ \cdot Cl \rightarrow Cl_2$$

In the higher alkyl halides the tendency is toward rearrangements of alkyl halides rather than to the production of free halides.

(b) Aromatic Compounds

Aromatic compounds are considerably more stable than are aliphatic compounds and alternative methods of energy dissipation are favoured in preference to dissociation. The stabilizing effect of the ring system also extends to attached side chains. Although relatively more stable than aliphatic compounds, aromatic compounds are not immune to the effects of radiation. Fission in the side chains does occur—particularly in the longer chains and the ring systems themselves do react to a certain degree. The irradiation product of benzene is a mixture of compounds containing chiefly C_{12} and C_{18} fractions.

One of the surprising features is the ability of aromatic compounds to extend their stability to other aliphatic compounds with which they are mixed by acting as an energy sink.

(c) Polymers

Irradiation of long chain polymers *e.g.* polyethylene, frequently results in cross-linking of the polymer chains. This produces an increase in the polymer molecular weight. With effectively one cross-link per polymer chain, a single piece of polymer becomes effectively one single molecule. There is consequently a change in the physical properties of the polymer, particularly in its increased resistance to deformation at elevated temperatures. Polyethylene, for example, which normally softens at about 90 °C and melts at 120 °C, can be taken to 250 °C after an irradiation corresponding to $2 \cdot 10^6$ rads.

8

Isotope Measurement and Separation Methods

Mass Spectrometry

The function of a mass spectrometer is to sort out atoms and molecules according to their mass. Most commercially manufactured instruments employ a magnetic field to split up a mass spectrum, in much the same way as a glass prism divides up the visible light spectrum. From the original "string and sealing wax" laboratory contraptions designed to separate and distinguish isotopes, modern instruments have been developed for the routine measurement of isotope ratios and for the determination of the masses of molecules and molecular fragments. It is in this latter field that the mass spectrometer finds extensive chemical application today.

Credit for the discovery of isotopes is shared by two people, Soddy and Aston. In the study of naturally occurring radioactive decay series during the early part of the century it was found that the element lead was the end product of both the uranium-238 and the thorium-232 decay series. This was indeed strange, since mass loss in radioactive decay was attributed to α-particle emission with the loss of four mass units each time. If the masses of the two parents differed by six mass units, and the only mass changes were in steps of four units, it would imply that the lead end product form one system could not be the same as the lead end product of the other. At the least it would have to differ by two mass units. Soddy confirmed the existence of two different varieties of lead when, in 1914, he showed by chemical means that the atomic weight of lead in thorium minerals was 208 mass units, which was one greater than that determined for "normal" lead. This was the first chemical evidence to confirm the existence of isotopes.

The previous year Thomson had extended his investigation of positive rays produced in gas discharges to a study of neon. He found that when the collimated positive rays produced in neon were subjected to crossed electric and magnetic fields, the beam divided and appeared in two different spots on the photographic plate at the end of the tube. One of the spots corresponded to a mass number of 20, while the other, which was only one-tenth as bright, corresponded to a mass number of 22. At first, he attributed the results to the presence of an impurity. He passed to his assistant, Aston, the task of purifying the neon. Chemical attempts failed to effect a separation, though Aston did achieve a partial change in the relative intensity of the two species by repeated diffusion of the gas through pipe-clay tubes. Two fractions obtained in this way were shown by Aston to be chemically identical and to have atomic weights, as determined from their densities of 20.15 and 20.28. Toward the end of 1913 Aston interpreted these results as being evidence for the existence of two different types of neon—a fact which Thomson was reluctant to accept for many years. Besides providing evidence for what we now know as isotopes, Aston's work also represented the first isotope enrichment process ever achieved; a process which was fundamentally the same as that employed during the second World War for uranium isotope separation.

The first World War interrupted research, but at its conclusion, Aston followed up his earlier work, and in 1919, completed the first mass spectro*graph* with which he not only verified that neon consisted of different isotopes, but also that eight of eighteen other elements, which he studied, were also composed of different isotopes. The mass spectrograph constructed by Aston, was in effect, a refinement of Thomson's positive ray apparatus and consisted of successive electric and magnetic fields to separate the isotopes (Fig. 8.1a).

During this same period, but on the other side of the Atlantic, Dempster designed and built the first mass spectro*meter* which employed electronic detection of the ion beam instead of a photographic plate as used by Aston. This was possible because of the higher ion currents which he obtained using only magnetic deflection and using the "lens" effect of the field of focus the ions form a less collimated source. This lens effect, which is employed in most modern instruments today, arises form the fact that all ions with the same momentum have the

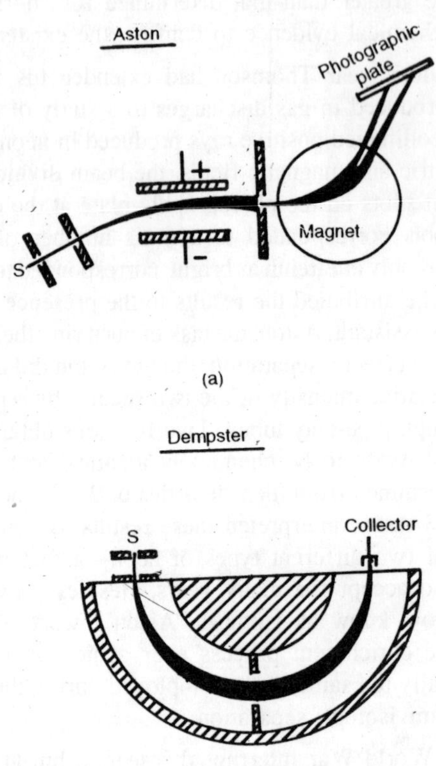

Fig. 8.1. (a) Aston's mass spectrograph. (b) Dempster's mass spectrometer.

same radius of curvature in a magnetic field. From this simple fact, it can be shown by geometric construction that a divergent ion beam entering the field from an "image" slit, on a line joining the slit and the apex of the magnet, will converge at a point on an extension of this same line on the other side of the magnet (c.*f*. Fig. 8.1b and Fig. 8.3).

To appreciate the principles of operation of the various types of mass spectrometer which have been evolved, it is first necessary to understand the behaviours of ions in electric and magnetic fields:

(a) Acceleration of an Ion in an Electric Field

An ion of mass m and charge e, originally at rest, will be accelerated through a potential difference V to, a velocity v given by:

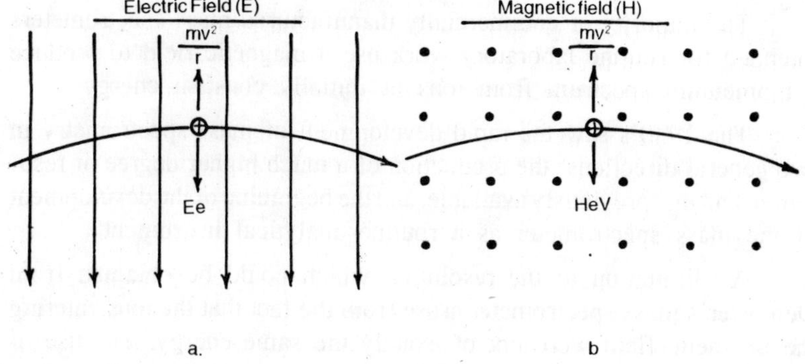

Fig. 8.2. Effect on a positively charged ion moving in : (a) an electric field; (b) a magnetic field.

Kinetic energy acquired = $\frac{1}{2}\, mv^2 = eV$ (1)

(b) Deflection of an Ion moving in a Transverse Electric Field

An ion moving at right angles to an electric field E (*N.B.E* is related to V by $E = V/d$, where d is the distance across which V is applied), will follow a curved path with a centripetal force

$$Ee = \frac{mv^2}{r} \quad \text{or} \quad r = \frac{mv^2}{Ee} \quad \text{(Fig. 8.2a)} \qquad (2)$$

From this it will be seen that for a transverse electric field the radius of curvature is a function of the *energy* of the ion.

(c) Deflection of an Ion moving across a Magnetic Field

An ion moving at right angles to a magnetic field will also be deflected— this time across the field (Fig. 8.2b). The force due to the magnetic field *Hev* being balanced by the centrifugal force.

$$Hev = \frac{mv^2}{r} \quad \text{or} \quad r = \frac{mv}{He} \qquad (3)$$

This time the radius of curvature is a function of the *momentum* of the ion.

Thus, an electric field acting in the same direction as the motion of an ion accelerates it, a transverse electric field separates ion according to their energy, and a magnetic field sorts them out according to their momentum.

The majority of commercially manufactured mass spectrometers intended for routine laboratory work use a magnetic field to produce a momentum spectrum from ions of initially constant energy.

The 1930's saw the rapid development of mass spectrometry in two general directions: the production of a much higher degree of resolution than that previously available, and the beginning of the development of the mass spectrometer as a routine analytical instrument.

A limitation to the resolution which could be obtained from Dempster's mass spectrometer arose from the fact that the ions entering the magnetic field were not of exactly the same energy. The use of an electric filed ahead of the magnetic field could, however, select ions of a single energy to be analysed by the magnetic field. A large number of instruments using successive electric and magnetic deflection were therefore evolved. These instruments have a mass resolution of the order of one part in ten thousand and are of considerable value in the determination of exact nuclide masses for the calculation of binding energies. These high resolution instruments are also of value in chemical analysis, since the high degree of mass resolution available makes it possible to distinguish between such ions as Ch_4^+ and O^+, both of which have a nominal mass number of 16.

Technical refinements in the design of magnetic deflection instruments and their associated electronics—particularly the work of Nier—have led to the wide scale adoption of the magnetic sector type of mass spectrometer for routine work in chemical laboratories. Originally developed by Nier as a 60° deflection instrument, it is now commonly used with either 60° or 90° deflection. Like Dempster's instrument it uses both momentum analysis and the focussing effect of the magnetic field.

The general arrangement is shown in Fig. 8.3.

Ions produced by electron bombardment in the source are accelerated by a potential V through one or more electrodes in the form of wide slits, then pass through two fine collimating slits, between which are a pair of beam deflection electrodes. The second collimating slit, which is only a few thousandths of an inch wide, acts as an "image" or virtual source of ions. The ions then proceed down the tube in a slightly divergent beam and enter the magnetic field H where they are deflected in an arc of radius r and emerge as a convergent beam to pass down the tube to the exit slit and ion current detector. The ion currents, which are of the order of one micro-micro ampere are measured by an electrometer amplifier.

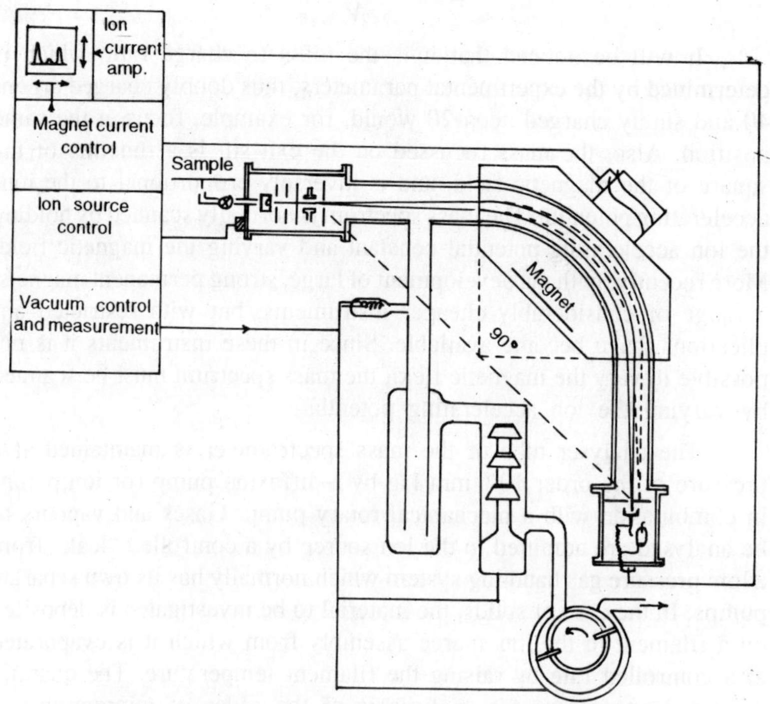

Fig. 8.3 Arrangement of a magnetic deflection mass spectrometer of the type used for many routing chemical application.

As mentioned earlier, the kinetic energy ($\frac{1}{2}mv^2$) acquired by the ions is equal to eV, *i.e.*

$$\frac{1}{2}\, mv^2 = eV$$

and the force on the ions due to the magnetic field (Hev) is balanced by the centrifugal force mv^2/r, *i.e.*

Eliminating v from the above equations, we obtain a relationship between the mass to charge ratio and the measurable parameters:

$$\frac{m}{e} = \frac{H^2 r^2}{2V}$$

N.B. If m is in a.m.u., e is the number of electronic charges on the ion, H is in gauss, V in volts and r in cm, then

$$\frac{m}{e} = \frac{4.79 \cdot 10^{-5} r^2 H^2}{V}$$

It will be noticed that it is the mass to charge ratio which is determined by the experimental parameters, thus doubly charged argon-40 and singly charged neon-20 would, for example, focus at the same position. Also, the mass focussed on the exit silt is a function of the square of the magnetic field, and is inversely proportional to the ion-accelerating potential. The mass spectrum is normally scanned by holding the ion accelerating potential constant and varying the magnetic field. More recently, with the development of large, strong permanent magnets, a range of considerably cheaper instruments, but with restricted applications, have become available. Since in these instruments it is not possible to vary the magnetic field, the mass spectrum must be scanned by varying the ion accelerating potential.

The analyser tube of the mass spectrometer is maintained at a pressure of the order 10^{-6} mm Hg by a diffusion pump (or ion pump) in combination with a mechanical rotary pump. Gases and vapours to be analysed are admitted to the ion source by a controlled "leak" from a low pressure gas handling system which normally has its own separate pumps. In the case of solids, the material to be investigated is deposited on a filament in the ion source assembly from which it is evaporated at a controlled rate by raising the filament temperature. The quantity of sample necessary for analysis is of the order of micrograms.

Most commercially manufactured magnetic deflection instruments have a radius of curvature of from six to twelve inches with corresponding resolutions of from one part in 300 to one part in 3000. Instruments which employ an electric filed ahead of the magnetic field have a resolution some tenfold greater than that which could be attained with the magnetic field alone. Thus there are commercially available instruments with a resolution able to meet almost all chemical applications.

Not all mass spectrometers which have been developed follow the somewhat traditional line of development outlined above. Typical of some of the more recent developments are the high resolution mass synchrometer develop by L.G. Smith in 1951, and Bennett's low resolution radiofrequency mass spectrometer which uses neither magnetic nor transverse electric fields to resolve the mass spectrum.

The mass synchrometer, as it is called, makes the greatest possible use of a circular magnetic field. In this instrument, shown in Fig. 8.4a,

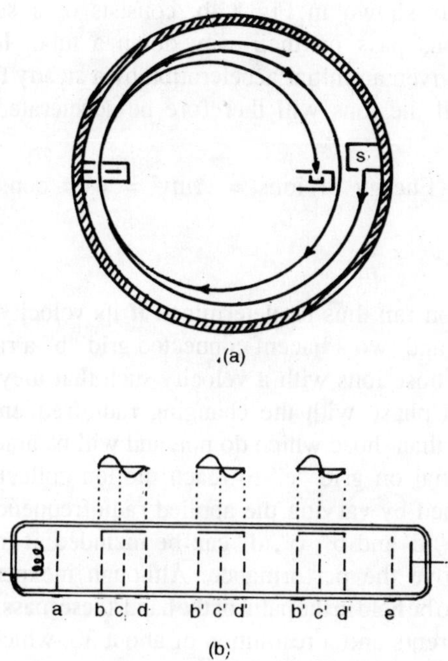

Fig. 8.4. (a) The mass synchrometer. (b) The radiofrequency mass spectrometer.

ions produced in the source are not accelerated by a steady potential, but by a pulse of short duration. A bunch of ions thus produced travel in a circular orbit under the influence of the magnetic field, and after half a revolution they reach a serics of three slits. Here they receive a small retarding pulse which reduced their energy and sends them on an inner orbit so that they do not collide with the rear of the source assembly. After several predetermined orbits they receive a third pulse as they pass through the slits which again changes their energy and brings them to the ion collector. Although electronically complex, this type of system is compact, makes greatest possible use of the field and has a mass resolution in excess of 10,000. In practice, several bunches of ions are simultaneously orbiting—the effect being to produce a modulated in beam and thereby avoiding extremely low mean ion currents.

Typical of instruments which do not employ magnetic deflection is the radiofrequency mass spectrometer, developed in 1946. This instrument, which is shown in Fig 8.4b, consists of a series of grids through which ions pass on their way down a tube. Ions produced in the source are given an initial acceleration by a steady fixed potential V at grid "a". All the ions will therefore be accelerated to the same energy (eV).

$$\text{Energy of ions} = \tfrac{1}{2}mv^2 = eV = \text{constant}$$

$$v^2 \propto \frac{1}{m}$$

The mass of an ion fan thus be determined if its velocity is measured. Between grid "c" and two adjacent connected grid "b" a radiofrequency field is applied. Those ions with a velocity such that they pass through the three grids in phase with the changing radiofrequency field, will gain more energy than those which do not, and will be able to overcome a retarding potential on grid "e" to reach the ion collector. The mass spectrum is scanned by varying the applied radiofrequency. Additional sets of grids b′, c′, d′ and b″, c″, d″ can be included at intervals down the tube to improve the performance. Although inexpensive, robust, and small enough to be held in the palm of the hand, these mass spectrometers have high ion currents and a resolution of about 30, which is sufficient for many applications. Their usefulness is also extended by the fact that they are capable of operating at higher gas pressures than the more conventional instruments.

Applications of the Mass Spectrometer

(a) Isotope measurement

The relative abundance of the isotopes of the naturally occurring elements is now largely a matter of record. However, in some fields, the measurement of isotope ratios is still important; for example

(i) In geochronology, where certain isotope ratio measurements can be used to determine the age of rock structures.

(ii) In inactive tracer analysis, where there are no suitable radioactive tracers available. Separated or enriched stable isotopes are then employed and subsequently measured with a mass spectrometer. This technique is particularly useful in biochemical and organic reaction mechanism studies using deuterium, nitrogen-15 and oxygen-18.

(iii) In isotope dilution analysis where a known quantity of a specific isotope is added to a sample containing the normal naturally occurring isotope ratio, the subsequent measurement of the isotope ratio, after through mixing, gives a measure of the total quantity of the element present. Because the method depends only upon the measurement of isotope ratios, quantitative recovery of the element under investigation is not necessary. The method is also particularly sensitive and can be used to determine trace quantities of an element in a sample down to concentrations of only 1 part in 10^{12}.

(iv) In the investigation of the phenomenon of biological fractionation, where certain organisms have been found to concentrate one or more isotopes in preference to other isotopes of the same element.

(b) "Cracking pattern". Organic analysis

Organic compounds, when bombarded by electrons in a mass spectrometer ion source, break up into ionized fragments. These ions provide a mass spectrum characteristic of the molecule bombarded and of the bombarding electron energy—the higher the energy, the more complex the spectrum becomes. For a given electron energy, the same molecule produces the same spectrum or *cracking pattern* as it is called. Hundreds of these are catalogued for cross reference and identification, and the method is widely used for routine analysis of gas mixtures in the petrochemicals industry.

This technique can also be applied to the elucidation of hitherto unknown organic structures by observing the appearance of spectrum peaks corresponding to particular groups *e.g.* hydroxyl, ethyl, etc. Closely related with this field of study is the measurement of *appearance potentials*. A gas or vapour in the ion source is bombarded by electrons of controlled variable energy to determine at what electron energies specific structural groups appear in the mass spectrum, and thereby to measure bond energies.

For example, it is known from thermodynamics that the energy of the reaction:

$$C_2H_6 + CH_4 \rightarrow C_3H_8 + 2H \text{ is } 5.1 \text{ eV} \quad (1)$$

The measured appearance potential for the $C_2H_5^+$ ion produced from ethane bombardment

$$C_2H_6 \rightarrow C_2H_5^+ + H \text{ is } 15.2 \text{ eV} \quad (2)$$

and the measured appearance potential for the $C_2H_5^+$ ion produced from propane bombardment

$$C_3H_8 \rightarrow C_2H_5^+ + CH_3 \text{ is } 14.5 \text{ eV} \qquad (3)$$

By adding (1) and (3) and subtracting (2), we get

$$CH_4 \rightarrow CH_3 + H \text{ and } E = 4.4 \text{ eV}.$$

(c) Precision Measurement of Nuclide Masses

High resolution mass spectrometers are used in the measurement of nuclide masses to high orders of precision in such physical studies as the determination of nuclear binding energies. Such application are, however, outside the scope of this text.

(d) Leak Detection

Low resolution mass spectrometers of the radiofrequency type, or small magnetic deflection instruments using permanent magnets, are often used in vacuum and gas handling pipework as leak detectors. These instruments are preset to record a specific mass—usually mass 4—and a gas such as helium is introduced into the system. A "sniffer" attached to the ion source by a flexible tube is then used to "sniff" around the outside of the equipment in search of a leak. The method is particularly sensitive and useful in the detection of very small leaks in industrial or laboratory vacuum systems.

Isotope Separation Methods

It is not intended to instruct the student in the technique of isotope separation: nevertheless, it is likely that he will have occasion to use commercially available separated isotopes. For this reason, as well as for general interest, a brief qualitative explanation of isotope methods is included.

Besides the more obvious applications such as the use of uranium-235 as a reactor fuel and heavy water as a moderator, separated isotopes find a variety of applications. These include such techniques as isotope dilution analysis and the use of separated stable isotopes as tracers in reaction mechanism studies, where suitable radioactive isotopes of the element under investigation do not exist.

Separated isotopes are also sometimes used as neutron targets to produce a single radioactive isotopic species where neutron irradiation of the naturally occurring mixture of isotopes would produce additional unwanted activities.

There are also some more specialised application such as, for example, the use of deuterium in ultraviolet lamps for spectrophotometers. The separation of isotopes of a large number of elements is an industrial undertaking and the separated isotopes are available commercially through normal isotope distribution agencies.

Isotope separation methods depend for their operation on either small physical (*i.e.* mass) differences, or on the slight chemical differences which exist between isotopes of some of the lighter elements. Chemical difference between isotopes is perhaps surprising, since isotopes are frequently defined as being chemically identical species which differ only in atomic mass. In fact, they would be better defined as being "almost chemically identical". Chemical properties depend upon the outer, valence, electrons of an atom. The ease whereby these electrons enter into chemical bonds depends upon the force of attraction exerted on them by the nucleus. This is almost exclusively—but not entirely—the electrostatic attraction of the protons in the nucleus. There is also a much smaller contribution in the form of a "gravitational" force exerted by the nucleus. In light elements the difference in the contribution of this gravitational force is greatest; for example, it is twice as great in deuterium as it is in hydrogen.

Separation processes usually depend on a very large number of stages each of which produces only a slight isotope enrichment. The *stage separation factor (S)* is defined as the ratio of the abundance of a given isotope in the enriched state as it leaves a stage to the ratio of its abundance as it enters. The overall separation factor for a process consisting of *n* stages in series is S^n. Although the stage separation factor is the parameter by which different methods can be compared, it alone does not define the overall efficiency of a process. For example, extremely high separation factors are achieved with electromagnetic separators (calutrons) but the recovery of the enriched isotope at the collector is only a few per cent of the initial charge.

Separation Methods

(a) Chemical exchange

As mentioned above, chemical properties are slightly dependent upon atomic mass. This is particularly so where the mass ratios of the isotopes to be separated are large, and as a consequence, chemical exchange methods on a commercial scale have only been really practicable up to nitrogen isotope separation. Of particular importance is the separation of hydrogen isotopes (hydrogen and deuterium).

In cases of compounds of hydrogen, if one is liquid and the other gaseous the deuterium tends to concentrate in the liquid phase. Two examples are the exchange between hydrogen and water, hydrogen sulphide and water. The equations representing these exchange reactions together with the separation factors at 25°C and 100°C are:

Reaction	$S_{25^\circ C}$	$S_{100^\circ C}$
$HD + H_2O \rightleftarrows H_2 + HDO$	3.87	2.69
$HDS + H_2O \rightleftarrows H_2S + HDO$	2.34	1.92

Exchange is effected by passing the gas through a long packed column containing the liquid phase (Fig. 8.5a). Although the hydrogen exchange reaction has the higher separation factor, it has two disad

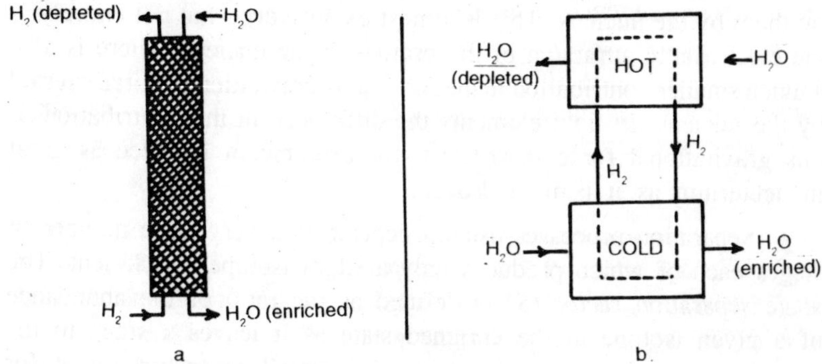

Fig. 8.5. Methods for the enrichment of hydrogen isotopes by chemical exchange.

vantages. These are:

(a) A catalyst is required.

(b) The regeneration of enriched hydrogen from the water for the next stage required electrolysis and is consequently expensive.

The former disadvantage does not exist in the H_2S system, while the latter objection can be overcome in both of the systems above by using what is known as the dual temperature process. In this process, equilibrium is established between hydrogen (or hydrogen sulphide) in two separate columns, which are maintained at different temperatures, the gas being passed first through one, then through the other and then back again (Fig. 8.5b). The equilibrium concentration of deuterium in the gas as it leaves the cold column is less than the equilibrium concentration in the hot column from which it abstracts more deuterium to transfer to the cold column on the next cycle. The separation factor for a single dual temperature stage is lower (= 3.87/2.69 = 1.44 for hydrogen exchange between 25° C and 100° C), but expensive electrolytic processes are avoided. The lower stage efficiency is overcome by using a large number of stages.

(b) Distillation

Different isotopes of the same element and their compounds have slightly different boiling points. The technique is limited to the lighter elements and their compounds, H_2O and D_2O for example, differ in boiling point by 1.4° C, D_2O being the higher. Distillation is used as an intermediate method in the commercial production of heavy water after initial enrichment by chemical exchange. The phenomenon of isotope fractionation is also sometimes observed to disadvantage in mass spectrometry. Lithium isotope ratios, for example, are frequently observed to change progressively as the lithium-6 evaporates preferentially from the solid source filament assembly.

(c) Electrolysis

Separation factors as high as 6 are possible for the separation of hydrogen isotopes by electrolysis of aqueous solutions—usually of potassium hydroxide. Early prewar processes for the production of limited quantities of heavy water for research purposes were based on electrolytic methods. These are now generally only employed as a finishing process in the production of heavy water. The electrolytic process which consists of several electrolysis stages in series, each with its own recombiner feeding back to the previous stage is shown in Fig. 8.6.

Before the second World War, deuterium was the only isotope which had been separated in any appreciable quantity. Most of the world supply—some 165 kg produced by the Norsak Hydro Company in Norway—was in the Paris laboratory of Joliot—Curie. This entire stock was smuggled out of France into England in June 1940 whilst the Germans were advancing on Paris and was used in much of the early work on the study of a possible uranium chain reaction.

(d) Gaseous diffusion

This method which was used by Aston on neon isotopes in 1913, is historically the oldest method whereby an isotope enrichment has

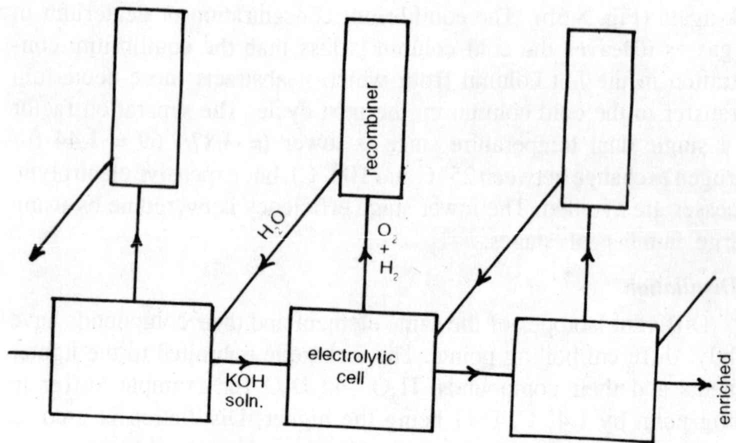

Fig. 8.6. Multistage series arrangement for the enrichment of hydrogen isotopes by electrolysis.

been effected. Little use was made of this method, however, until the war- time Manhattan Project when it was one of the methods investigated for the production of kilogram quantities of uranium-235 from natural uranium. It is based on the fact that the rate of diffusion of gas through a porous membrane is inversely proportional to the square root of its density and hence to its molecular weight. The only suitable volatile compound of uranium is uranium hexafluoride (UF_6) with a boiling point at atmospheric pressure of 56° C. The theoretical stage separation factor is given by:

$$S = \sqrt{\frac{238 + (6 \times 19)}{235 + (6 \times 19)}} = 1.0043$$

In practice only about one-third of the theoretical separation factor is achieved, and some thousand stages, each with its own pumps to maintain a pressure gradient across a diffusion barrier, are necessary to produce 99% pure $^{235}UF_6$.

Although stage separation factors are low, the method proved to be the best for large scale continuous operation. All uranium-235 produced at present is separated by gaseous diffusion of UF_6.

(e) Thermal diffusion

When a temperature gradient is established in a gas containing two or more isotopes there is an extremely small tendency for the isotopes to separate—one to concentrate in the hot region and the other in the cold. This tendency is so slight that the method would be useless from any practical point of view, were it not for the work of Clusius and Dickel in Germany in 1938. They carried out a successful isotope enrichment in a very long vertical tube, at the axis of which was a wire heated electrically to about 600°C. The heated wire in the tube produced two effects:

(i) Slight thermal diffusion radially.

(ii) Convection currents vertically with the hot gas rising in the centre and cooler gas descending at the outer surface.

Although there existed an infinitesimally small enrichment effect radially, the counter current flow, which was established vertically, had the effect of producing and extremely large number of stages vertically and considerably multiplied the concentration effect. This method is particularly suitable for small scale laboratory enrichments of many isotopes, since the equipment—shown diagrammatically in Fig. 8.7 is simple, inexpensive and has no moving parts.

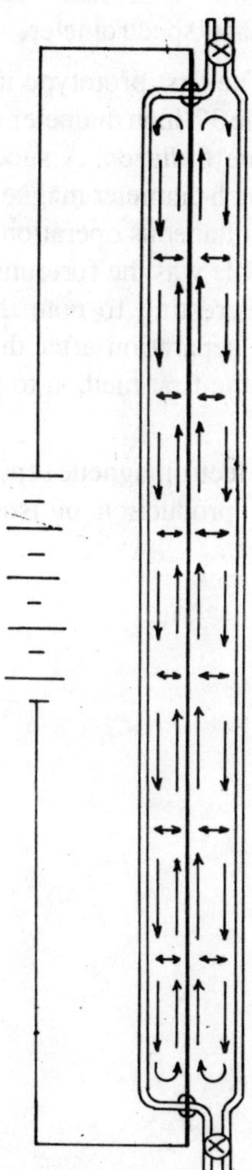

Fig. 8.7 : The Clusius-Dickel column used to effect isotope enrichment by thermal diffusion.

(f) Gas Centrifuge

This apparatus consists of a vertical cylinder which is rotated at an extremely high velocity. There is a very slight concentration of the heavier isotopes at the outer surface with the lighter ones at the centre. By setting up a counter current vertical flow, a situation is established similar to that described above for the thermal diffusion method. The gas centrifuge method was operated successfully on a pilot plant scale to separate uranium isotopes during the war, but although good isotope enrichment was achieved, mechanical difficulties associated with large scale continuous operation, led to its abandonment in favour of gaseous diffusion.

(g) Electromagnetic Separators (Calutron)

These instruments are essentially a mass spectrometer on a grand scale. They rely on the production and acceleration of ions in an ion source and their subsequent separation in a magnetic field—usually using 180° deflection—very similar to a giant version of Dempster's first mass spectrometer.

The first prototype instrument was constructed at the end of 1941 using the 37-inch diameter *C*alifornia *U*niversity cyclo*tron* magnet, hence the name *Calutron*. A successor was built some six months later using a 184-inch diameter magnet with a pole gap of 72 inches, which permitted the simultaneous operation of several ion beams within the one magnetic field. This was the forerunner of many large electromagnetic separators. It is interesting to note that although work began on this method of isotope separation after the centrifuge and gaseous diffusion methods, it was the first method to produce large amounts os separated uranium-235.

Electromagnetic separators are at present operated in many countries for the production of isotopes of a very large number of elements.

9
Charged Particle Accelerators, Neutron Sources, Production and the Actinides

The various types of nuclear reaction which result from the bombardment of nuclei by charged particles and neutrons were discussed in chapter 3. This Chapter describes the methods employed for the production of bombarding particles. These methods will be divided into two groups: those for the production of charged particles, and those for the production of neutrons. Because the production of the actinide elements is closely associated with the operation of the nuclear reactor, it is convenient to discuss the properties of these elements at this stage. This Chapter is therefore divided into three section:

(1) Charged particle accelerators,

(2) Neutron sources, and

(3) The actinide elements.

Charged Particle Accelerators

An obvious method for the acceleration of charged particles, such as hydrogen ions (protons), is by the application of an electric field. Particles carrying an electric charge (e) will be accelerated by the application of a potential (V) to a kinetic energy given by:

$$\text{K.E.} \; (= \tfrac{1}{2}mv^2) = eV$$

This units used to express the energy of these particles are the electron volt (eV), and its multiples the kiloelectron volt (keV), and million electron volt (MeV). These correspond to the energies which would be acquired by a charged particle carrying unit electronic charge when it is accelerated through a potential difference of one volt, 10^3 volts and 10^6 volts respectively (*N.B.* 1 MeV = 1.6 · 10^{-6} erg).

Potentials of several tens of thousands of volts can be readily obtained from conventional transformer—rectifier circuits. However, the potentials necessary to accelerate light ions such as hydrogen and helium nuclei, to an energy sufficient to overcome the potential barrier of even light nuclei *e.g.* lithium and beryllium are of the order of one MeV. For heavier nuclei which carry a greater nuclear charge, the energy required is correspondingly higher. Special methods for the acceleration of particles to these energies are therefore necessary, *e.g.* the Cockroft—Wolton voltage multiplier, the Van de Graaff generator, the linear accelerator and the cyclotron.

(A) The Cockroft—Walton Voltage Multiplier

Working in Rutherfords laboratory in 1931, Cockroft and Walton achieved the first artificial transmutation when they bombarded lithium with high speed protons.

$$^{7}_{3}Li + ^{1}_{1}H \rightarrow ^{4}_{2}He + ^{4}_{2}He$$

The protons were accelerated in an evacuated tube by the application of a very high potential derived from a voltage multiplier circuit (Fig. 9.1a). This device uses a single high voltage transformer in conjunction with a large number of rectifiers and condensers. Its operations is best understood by considering the first cycle of the transformer, when the right hand side is at a negative potential. Current will flow through D_1 to charge C_1. On the reverse half of the cycle, when the right hand side is positive, no current can flow back through D_1, but part of the charge in C_1 will flow through D_2 to begin charging C_2. On the second cycle when the right hand side is negative, more charge will be passed through D_1 to charge C_1 and some of the charge on C_2 will be passed through D_3 to C_3. As the process is repeated, the charge on each condenser progressively builds up until each is charged to a potential approximately equal to the output voltage of the transformer. The output voltage of the transformer is thereby multiplied several times over; hence the name *voltage multiplier* which is applied to the circuit. If the output potential of the transformer shown in Fig. 9.1a is 25,000 volts, then the overall output of the circuit shown would be approximately 150,000 volts.

The Cockroft—Walton voltage multiplier has the advantages of simplicity and the production of fairly large ion currents at a steady potential. It is widely used for the production of accelerating potentials up to 10^6 volts, at which insulation problems set a practical limit. It

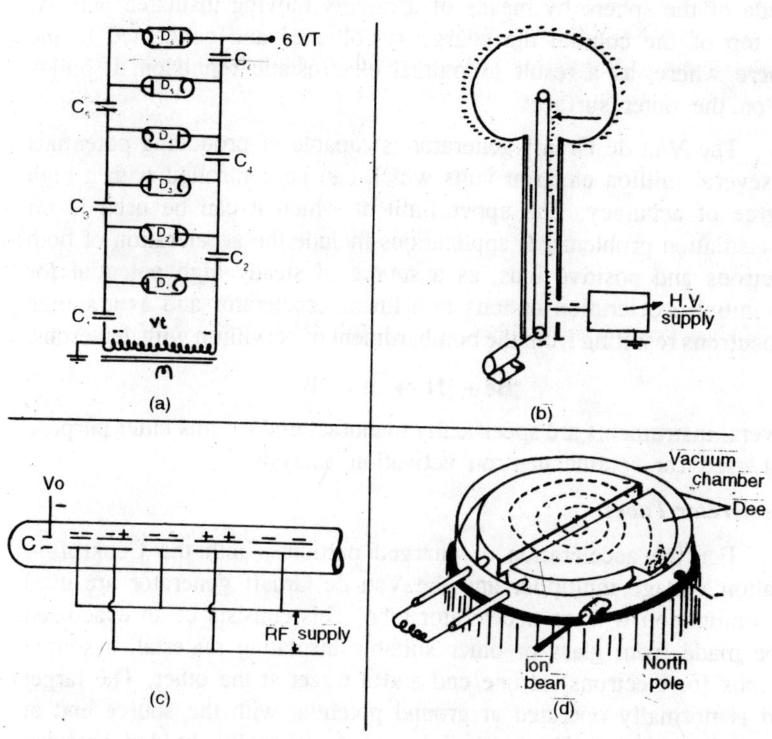

Fig. 9.1. Schematic diagrams of : (a) Cockroft—Walton voltage multiplier; (b) Van be Graaff generator; (c) linear accelerator; (d) cyclotron (with top of vacuum chamber and south pole removed).

is particularly suitable for the production of neutrons by the bombardment of tritium adsorbed on titanium or zirconium targets by deuterons of relatively low energies (~ 0.2 MeV).

(B) The Van de Graaff Generator

The Van de Graaff generator, which bears the name of its inventor, was developed at the Massachusetts Institute of Technology in 1931. It is an electrostatic generator, the basic principle of which is similar to the more familiar Wimshurst machine. A high potential is built up on a large hollow conducting sphere at the top of an insulated column (Fig. 9.1b). Electric charge is conveyed from a direct current voltage supply of some 10 to 20 kilovolts at the base of the column to the inside of the sphere by means of a rapidly moving insulated belt. At the top of the column the charge is collected and conducted to the sphere where, as a result of mutual electrostatic repulsion, it builds up on the outer surface.

The Van de Graaff generator is capable of producing potentials of several million electron volts which can be controlled with a high degree of accuracy. The upper limit at which it can be used is set by insulation problems. It applications include the acceleration of both electrons and positive ions, as a source of steady high potential for the initial acceleration of ions in a linear accelerator and as a source of neutrons resulting from the bombardment of beryllium with deuterons.

$$^{9}_{4}Be + ^{2}_{1}H \rightarrow ^{1}_{0}n + ^{10}_{5}B$$

Several instruments are specifically manufactured for this latter purpose and used for routine neutron activation analysis.

Accelerator Tubes

For the acceleration of charged particles, both the Cockroft—Walton voltage multiplier and the Van de Graaff generator are used in conjunction with an accelerator tube. This consists of an evacuated tube made from glass or other suitable insulating material, a source of ions (or electrons) at one end and a target at the other. The target end is normally operated at ground potential with the source end at the high potential. The applied potential is usually divided between intermediate tubular accelerating electrodes spaced along the axis of the tube. These help to define the trajectory of the beam of charged particles and provide a focussing action.

(C) Linear Accelerator

The linear accelerator was first developed by Lawrence and Sloan in the United States during 1931. It is a device whereby the energy of ions is progressively increased by multiple acceleration. The linear accelerator consists of a series of hollow cylindrical electrodes called *drift tubes* which are arranged along the axis of an evacuated tube. Each successive drift tube is longer than the preceding one alternate tubes are connected in parallel, and the output from a high frequency, high power, oscillator is applied to the two groups of tubes (Fig. 9.1c).

A bunch of positive ions approaching the first drift tube, while it is at a high negative potential are accelerated towards the tube and into the field region inside. Here they "drift" through at a constant velocity unaffected by electric fields. If their velocity is such that they traverse the length of the tube during half of the period of the radiofrequency oscillation, they will arrive at the gap between the first and second tubes when the second tube is at its peak negative potential with respect to the first tube. They will be further accelerated, pass through the second drift tube at an increased velocity and, if the second drift tube is proportionately longer than the first, they will emerge in time to receive a further acceleration towards the third drift tube, and so on.

If V_0 is the initial fixed potential used to accelerate ions into system, and V is the potential difference between accelerating stages, then ions will have a velocity after the first stage corresponding to an acceleration through $V_0 + V$ volts. After the second stage they will have a velocity corresponding to an acceleration through $V_0 + 2V$ volts, and after n stages they will have a velocity corresponding to an acceleration through $V_0 + nV$ volts.

Energy after n stages = $\frac{1}{2} mv^2 = (V_0 + nV)e$ and the final velocity after n stages = $v = \sqrt{2e / m(V_0 + nV)}$ where m and e are the mass and charge of the ion accelerated.

In early experiments, Lawrence and Sloan used 21 drift tubes (overall length 80 cm) and a 3.5 megacycle supply to obtain mercury ions with an energy equivalent to an acceleration through 130,000 volts. In a later larger version, the energy of the mercury ions was increased almost ten-fold. Because of their large nuclear charge, and therefore their low probability of nuclear penetration, these ions were unsuitable for producing nuclear bombardment reactions. The acceleration of light ions, which have a higher velocity associated with the same kinetic energy, required much longer drift tubes or, alternatively, much higher frequencies which were not at the time available at high power levels. These factors led to the virtual abandonment of the linear accelerator in favour of the cyclotron until after the second World War when high frequency, high power oscillators developed for radar operation became available. Linear accelerators have since been developed for accelerating particles to extremely high energies (~ 1 BeV).

(D) The Cyclotron

This instrument was also developed by Lawrence in the United States during the early 1930's, and became the most widely used method for the acceleration of charged particles. The cyclotron also uses multiple acceleration of ions and in many respects it can be thought of as a linear accelerator which has been wrapped around itself in a spiral. The drift tubes are replaced by two flat hollow semicircular electrodes called *dees* because of their resemblance to the letter “D”. These dees, which are connected to a radiofrequency oscillator, are contained in a large flat vacuum tank between the poles of an electromagnet (Fig. 9. 1d).

Ions formed at the centre are attracted by the applied electric field into one of the dees where they drift in an electrically field free region. They are confined to a semicircular path by the magnetic field and, if they emerge from the dee after a period of time corresponding to half of the period of the oscillation, they will be accelerated across the gap between the dees and into the second dee where they will follow another semicircular path. This process is repeated many times, the ions following orbits of increasing radius after each successive acceleration until they reach the perimeter. Here they are deflected from their circular path by a negatively charged electrode and emerge through a thin aluminium widow.

The operation of the cyclotron depends upon the ions taking exactly the same interval of time; *i.e.* half of the period of the applied oscillation, on each semicircular orbit. The forces acting to constrain an ion in a semicircular path are

$$Hev = \frac{mv^2}{r}$$

Substituting ω =(the angular velocity) for v/r we obtain

$$\omega = \frac{He}{m}$$

It will therefore be seen that providing m is constant, the angular velocity is constant, and independent of the radius of the path followed. That is to say, that although the ions traverse larger semicircles as their energy increases, they still take the same time to reach the gap between the dees. The relativistic increase in mass of ions as they reach high velocities results in ions arriving at the gap between the dees out of phase and thereby places an upper limit of about 20 MeV, in the case of deuterons, on the energies which can be obtained by this method of particle acceleration.

Another important feature of the cyclotron is that ions do not have to reach the gap between the dees at the precise moment when the potential difference between the dees is at its maximum value. Those which reach the gap at a time behind the peak potential receive a smaller acceleration, but still take the same time to traverse their semicircular path as do the other ions. They thus arrive at the gap with the same time lag and eventually, after a greater number of orbits, reach the same energy as the ions that received maximum acceleration each time.

For resonance acceleration to occur the angular velocity ω must be such that one semicircle is traversed during half the period of oscillation:

$$\frac{\pi}{\omega} = \frac{T}{2}$$

or the frequency $f = \dfrac{1}{T} = \dfrac{\omega}{2\pi} = \dfrac{He}{2\pi m}$

It will be noticed that for a specific magnetic field strength the frequency is a function of *e/m*, thus for example, both deuterons and helium nuclei would require the same frequency. They would both the accelerated to the same final velocity although the lighter deuterons would have only half the energy of the helium nuclei.

To determine the maximum energy to which an ion can be accelerated, we must consider the forces acting on the ion in it outermost orbit:

$$Hev = \frac{mv^2}{R}$$

where R is the radius of the outermost orbit.

If we square and rearrange the expression, we obtain:

$$\tfrac{1}{2}mv^2 (= \text{kinetic energy}) = \frac{H^2e^2R^2}{2m}$$

It will be seen that the energy which can be obtained is a function of the radius, *i.e.* the size, of the cyclotron, and it is therefore common to find cyclotrons described in terms of their diameter.

Neutron Sources

There are three different types of neutron source available. These are :

(A) Radioactive neutron sources,

(B) Particle accelerator sources, and

(C) The nuclear reactor.

(A) Radioactive Neutron Sources

There are no neutron emitting radioactive nuclides which can be used directly as a source of neutrons. There are, however, several suitable α and γ emitting nuclides which can be used in conjunction with a suitable light element, such as beryllium, to produce neutrons by either an α, n reaction or by α γ, n reaction:

$$^{9}_{4}Be + ^{4}_{2}He \rightarrow ^{1}_{0}n + ^{12}_{6}C$$

$$^{9}_{4}Be + \gamma \rightarrow ^{1}_{0}n + ^{8}_{4}Be^{*}$$
$$\downarrow$$
$$2^{4}_{2}He$$

The neutron output from the majority of commercially available sources lies in the range 10^4 to 10^7 neutrons per second and this type of source represents a relatively cheap low flux source of neutrons. Comparative data for some of the more commonly used types are given in Table 9.1. With the exception of the antimony/beryllium sources, the sources are fabricated using radioactive nuclides. In the case of the antimony sources, inactive antimony is used to fabricate the source which is then subsequently activated by irradiation in a nuclear reactor to produce antimony-124 from the naturally occurring stable antimony-123.

Table 9.1

Radioactive Neutron Sources

Type of source	*Reaction*	*Half-life*	*γ dose-rate per 10⁶ n/sec (milli-roentgens/h at 1 metre)*	*Approx. cost in £ stg. per*	
				10^5 n/sec	10^7 n/sec
Actinum-227/beryllium	α, n	22 yr	8	120	—
Americium-241/beryllium	α, n	458 yr	1	100	1500
Lead-210/beryllium	α, n	22 yr	9	150	—
Polonium-210/beryllium	α, n	138 days	< 0.1	40	240
Radium-226/beryllium	α, n	1620 yr	60	100	—
Thorium-228/beryllium	α, n	1.9 yr	30	—	1000
Antimony-124/beryllium	γ, n	60 days	980	—	100

Desirable features are a long half-life and low γ dose rate. However, these features are usually associated with high costs, and the low initial cost of antimony/beryllium sources is, for example, offset by high shielding costs and the necessity for frequent reactivation.

(B) Particle Accelerator Sources

Particle accelerators may also be used as a source of high speed particles for the bombardment of light nuclei to yield neutrons. The most commonly used instruments for this purpose are the Van de Graaff generator and the Cockroft—Walton voltage multiplier, both of these are manufactured commercially for this specific application. Although the capital cost of these instruments is considerably higher than that of radioactive sources, the neutron output is also considerably greater (~ 10^8 to 10^{10} neutrons per second) and they are highly suitable for neutron activation analysis.

The bombardment of beryllium with 1 MeV deuterons from a Van de Graaff source yields approximately 10^8 neutrons per second per microampere of ion current, with considerably higher yields energies.

$$^{9}_{4}\text{Be} + {}^{2}_{1}\text{H} \rightarrow {}^{1}_{0}\text{n} + {}^{10}_{5}\text{B} \quad (Q = +3.79 \text{ MeV})$$

The reaction which is exoergic produces neutrons with an energy spectrum ranging up to 4.5 MeV (for 1 MeV bombarding deuterons). When a low energy neutron spectrum is required, and protons with an energy in excess of 1.88 MeV are available, the endoergic reaction between protons and lithium-7 can be employed

$$^{7}_{3}\text{Li} + {}^{1}_{1}\text{H} \rightarrow {}^{1}_{0}\text{n} + {}^{7}_{4}\text{Be} \quad (Q = -1.62 \text{ MeV})$$

Of importance where lower accelerating potentials are available (~200 keV), such as those from a Cockroft—Walton set, is the deuterium—deuterium reaction, and more particularly, the deuterium—tritium reaction

$$^{2}_{1}\text{H} + {}^{2}_{1}\text{H} \rightarrow {}^{1}_{0}\text{n} + {}^{3}_{2}\text{He} \quad (Q = +\ 3.27 \text{ MeV})$$

$$^{3}_{1}\text{H} + {}^{2}_{1}\text{H} \rightarrow {}^{1}_{0}\text{n} + {}^{4}_{2}\text{He} \quad (Q = +\ 17.6 \text{ MeV})$$

Deuterium targets have been available for a considerable time but it has only been in more recent years that tritium, a radioactive isotope of Hydrogen ($t_{1/2}$ = 12 yr), has become available in quantity. Tritium, which has a strong resonance absorption at deuteron energies slightly above 100 keV, is available as targets absorbed on titanium or zirconium. Costing approximately £40, and giving ~ 10^7 n/sec/μ ampere with a life of 40 hours at a beam current of 300 μ ampere, these targets are the basis of many moderately priced particle accelerator types of neutron source.

(C) The Nuclear Reactor

Toward the end of 1938, Hahn and Strassmann showed that when uranium nuclei were bombarded with neutrons they broke into two roughly equal fragments, or as we say, they underwent *fission*. Subsequently it was shown that more neutrons were produced in the process than were required to initiate it, and that a chain reaction should be feasible—neutrons liberated in one fission reaction providing the means to effect a second reaction and so on.

Subsequent investigations, notably in the United States, which were motivated by potential military applications led to a vast amount of fundamental information, to the discovery of new chemical elements, to the first controlled nuclear chain reaction at Chicago on December 2, 1942, and to the detonation of the first atomic bomb in New Mexico on July 16, 1945.

The three important fissile nuclides used as reactor fuels are: uranium-235 which occurs in nature and constitutes 0.7% of natural uranium, uranium-233 which is formed by neutron capture in thorium, and plutonium- 239 which results from neutron capture in uranium-238 :

$$^{232}_{90}Th \xrightarrow{n} {}^{233}_{90}Th \xrightarrow[23.3\ min]{\beta} {}^{233}_{91}Pa \xrightarrow[27.4\ d]{\beta} {}^{233}_{92}U \xrightarrow[1.62 \cdot 10^5\ yr]{\alpha}$$

$$^{238}_{92}U \xrightarrow{n} {}^{239}_{92}U \xrightarrow[23.5\ min]{\beta} {}^{239}_{93}Np \xrightarrow[2.33\ d]{\beta} {}^{239}_{94}Pu \xrightarrow[24.360\ yr]{\alpha}$$

Each of these three nuclides has a high cross-section for the absorption of low energy (0.025 eV) "*thermal*" neutrons and gives rise to between two and three neutrons per fission (Table 9.2).

Table 9.2

Nuclear Properties of Fissile Nuclides

Nuclide	*Fission cross-section (barns)*	*Average number of neutrons produced per fission*
$^{233}_{92}U$	525	2.55
$^{235}_{92}U$	580	2.47
$^{239}_{94}Pu$	740	2.91

In order to achieve a self sustained chain reaction, it is necessary to ensure that there are as many neutrons available to initiate each successive around of fission reactions as were used up in the preceeding round. That is to say, it is necessary to aim for a maximum neutron production together with minimum losses due to non-fissile absorption or by escape from the surface. Neutrons are produced in the *volume* of the fissile medium and lost form the medium through the *surface.* Therefore, if we consider a spherical system, neutron production will be a function of the cube of the radius while neutron losses from the surface will be a function of the square of the radius. That is to say, if we were to double the radius of a system, we would increase neutron production eight-fold and neutron losses from the surface only four-fold. As a consequence there is a certain minimum size in which a chain reaction can be attained—this is known as the *critical size.*

At this stage it is instructive to consider the factors associated with the establishment of chain reaction using natural uranium. As mentioned above, the three fissile nuclides exhibit high fission cross-sections for low energy neutrons. Unfortunately the neutrons which are released during fission have an average energy of about 2 MeV at which value the fission cross-sections are considerably reduced—that of uranium-235 being approximately one barn. Uranium-238 which constitutes 99.3% of natural uranium undergoes "fast fission" for neutron energies in excess of 1.5 MeV. The cross-section however, has a maximum value of only 0.5 barn at 2 MeV and above. The overall average fission cross-section for natural uranium with respect to initially released fission neutrons is therefore only about 0.5 barn and the greater majority of neutrons undergo non-capture scattering collisions, and soon decrease their energy below the uranium-238 fission threshold. In this intermediate energy region where the possibilities of either uranium-235 fission or uranium-238 capture exist, the latter would effectively mop up the greater majority of the neutrons because in this region uranium-235 has a fission cross-section of less than 100, barns and the uranium-238 has several resonance capture bands with cross-sections in as high as 10^4 barns. In addition, uranium-235 constitutes only 0.7% of the total uranium. A chain reaction could not, therefore, ever be attained in a piece of chemically pure natural uranium no matter what its size. There are, however, two ways in which a chain reaction can be established:

(i) Remove the uranium-238 by an isotope separation process. This is the basis of the fission bomb and the fast reactor.

(ii) Slow the neutrons to velocities below the resonance absorption region in the absence of uranium-238. This is the basis of the thermal reactor which has become a common source of power and isotope production.

The thermal reactor consists of an assembly of uranium fuel elements dispersed throughout a medium of low atomic number and low neutron capture and cross-section called *moderator*. The low atomic number is essential to ensure that the greatest possible amount of neutron energy is lost per collision, and the low cross-section is essential to reduce neutron losses. Graphite and heavy water are moderators in general use with natural uranium. In the case of enriched reactors ordinary water may be used. Beryllium is also highly suitable although its cost restricts its use.

In the region of thermal neutron energy the fission cross-section for uranium-235 in 580 brans and the capture cross section, leading to uranium-236, is 107 barns. Thus about one sixth of the thermal neutrons absorbed in uranium-235 do not produce fission and although an average of 2.47 neutrons are produced per fission only

$$2.47 \times \frac{580}{580+107} = 2.08 \text{ neutrons}$$

are produced per thermal neutron absorbed in uranium-235. In the case of natural uranium, the effective contribution of the uranium-235 cross-section is only 4.12 barns for fission and 0.76 barns for capture, while the effective uranium-238 capture cross-section is 2.75 barns. The number of fission neutrons produced for each thermal neutron captured in natural uranium is therefore

$$2.5 \times \frac{4.12}{4.12+0.76+2.75} = 1.35 .$$

Of these, one is essential to maintain the chain reaction and not more than the 0.35 can be lost in the slowing down process as a result of leakage, uranium resonance absorption, and capture by the moderator, materials of construction, control rods and samples inserted for irradiation. The thermal neutrons absorbed by the uranium-238, as well as those captured during the slowing down process are not wasted, for they lead to the production of fission plutonium-239 which can be recovered from the spent fuel elements by chemical processing methods. For every four atoms of uranium-235 which undergo fission, about three atoms of plutonium-239 will be produced in a natural uranium reactor. Long term irradiation of natural uranium can also lead to other higher actinides as will be seen in the next section.

Nuclear reactors provide neutron fluxes of up to 10^{15} neutrons per cm^2 per second. Besides providing a higher flux than is attainable by other methods, the flux is available over an infinitely greater volume permitting simultaneous irradiation of numerous samples.

THE ACTINIDE ELEMENTS

(A) Occurrence and Preparation

The actinide elements begin with actinium, atomic number 89, and end with Lawrencium, atomic number 103. All isotopes of all of these elements are radioactive. Only two of the elements, thorium and uranium, occur to any appreciable extent in nature. Small amounts of actinium and protactinium are present in uranium ores as decay production in the uranium-235 decay chain

$$^{235}_{92}U \xrightarrow[7.13\cdot 10^{8}\ yr]{\alpha} {}^{231}_{90}Th \xrightarrow[25.6\ h]{\beta} {}^{231}_{91}Pa \xrightarrow[32480\ yr]{\alpha} {}^{227}_{89}Ac \xrightarrow[21.2\ yr]{\beta}$$

Minute quantities of neptunium and plutonium resulting from natural neutron capture in uranium have been detected in uranium ores as a consequence of a deliberate search for these elements several years after their artificial production. The amount of plutonium present is of the order of one part in 10^{12} parts of uranium.

The only source of the majority of the actinides is their artificial production form a variety of bombardment reactions. Plutonium was the first artificial element ever to be produced in visible amounts. Literally tons of this element have since been manufactured in nuclear reactors which use natural uranium or low enrichment uranium as fuel. Long term irradiation of these fuels results in further neutron capture and leads to the elements americium (Fig. 10.2).

It will be noted in Fig. 9.2 that neptunium is produced as an

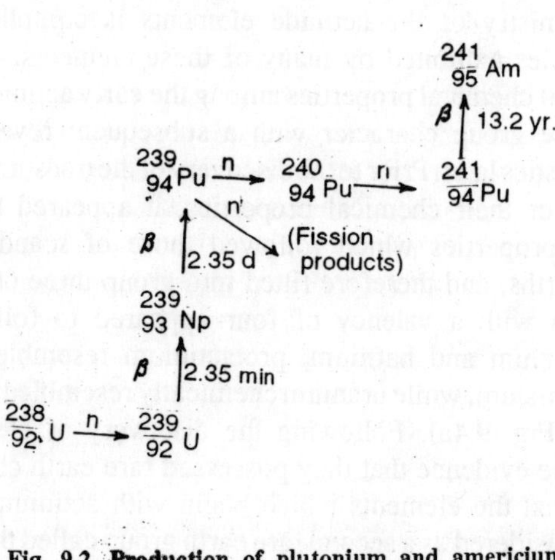

$$^{238}_{92}U \xrightarrow{n} {}^{239}_{92}U \xrightarrow[2.35\ min]{\beta} {}^{239}_{93}Np \xrightarrow[2.35\ d]{\beta} {}^{239}_{94}Pu \xrightarrow{n} {}^{240}_{94}Pu \xrightarrow{n} {}^{241}_{94}Pu \xrightarrow[13.2\ yr.]{\beta} {}^{241}_{95}Am$$

$$^{239}_{94}Pu \xrightarrow{n'} \text{(Fission products)}$$

Fig. 9.2. Production of plutonium and americium as a result of neutron irradiation of uranium.

intermediate. This particular isotope of neptunium has a half-life which is too short to be of very much value for chemical studies. A much longer-lived isotope of neptunium with a half-life of 2.14 · 10^6 years is, however, available from the spend fuel elements of highly enriched uranium-235 reactors (Fig. 9.3).

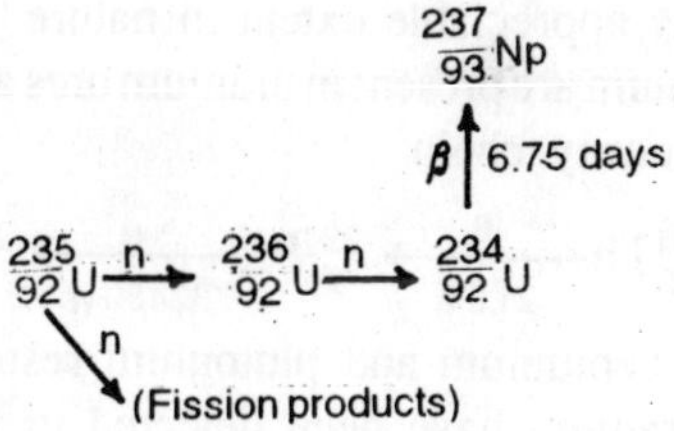

$$^{235}_{92}U \xrightarrow{n} {}^{236}_{92}U \xrightarrow{n} {}^{234}_{92}U$$

$$^{234}_{92}U \xrightarrow[\text{6.75 days}]{\beta} {}^{237}_{93}Np$$

$$^{235}_{92}U \xrightarrow{n} \text{(Fission products)}$$

Fig. 9.3. Production of neptunium-237 in enriched reactor fuels.

Long term irradiations of separated plutonium in high flux research reactors have produced, by a succession of neutron capture and β decay processes, quantities of the higher actinides ranging from grams of americium and curium (mainly curium-244) to sub-microgram amount of Einsteinium and Fermium. Although the forementioned actinide elements have been produced in quantity by neutron irradiation, their original discovery was in most instances as a result of particle accelerator bombardments. A summary of their discovery and the origin of the names of the "synthetic" actinide elements is given in Table. 9.3.

(B) General Chemical Properties of the Actinide Elements

The chemistry of the actinide elements is complicated by the multiple valencies exhibited by many of these elements. There is also a trend to exhibit chemical properties among the early actinides following a periodic table group character with a subsequent reversion to rare earth characteristics later. Prior to the discovery of the transuranic elements, and the study of their chemical properties, it appeared that actinium had chemical properties which followed those of scandium, yttrium and the rare earths, and therefore fitted into group three of the periodic table. Thorium with a valency of four appeared to follow on from titanium, zirconium and hafnium, protactinium resembled vanadium, niobium and tantalum, while uranium chemically resembled molybdenum and tungsten (Fig. 9.4a). Following the discovery of elements above uranium and the evidence that they possessed rare earth characteristics, it was found that the elements which begin with actinium could more correctly be considered as a second rare earth group called the "actinides". (Fig. 94b).

Table 9.3

Summary of the Discovery of the "Synthetic" Actinide Elements

Atomic number	*Name and symbol*	*Origin of Name*	*Method of orig. preparation*	*Date*
93	Neptunium Np	After the planet Neptune	Neutron irradiaon of uranium	1940
94	Plutonium Pu	After the planet Pluto	Deutron bombardment of uranium	1940
95	Americium Am	The Americas (analogue of Europium in the rare earths)	Neutron irradiation of plutonium	1944
96	Curium Cm	Honour of the Curies	Helium ion bombardment of plutonium	1944
97	Berkelium Bk	City of Berkeley, Calif.	Helium ion bombardment of americium	1949
98	Californium Cf	California University & State	Helium ion bombardment of curium	1950
99	Einsteinium Es	Honour of Einstein	Multiple neutron capture by uranium in a thermonuclear explosion	1952
100	Fermium Fm	Honour of Fermi	Multiple neutron capture by uranium in a thermonuclear explosion	1952
101	Mendelevium Md	Honour of Mendeleev	Helium ion bombardment of einsteinium	1955
102	Nobelium No	Honour of Nobel	*	(1957) 1958*
103	Lawrencium Lw	Honour of Lawrence	Boron ion bombardment of californium	1961

* A claim was made, by a joint U.K., U.S. and Swedish group in 1957, to have prepared an isotope emitting an 8.5 MeV α-particle by bombarding ^{244}Cm with ^{13}C ions. This was later disputed by another U.S. group, who subsequently claimed (1958) to have produced a 7.43 MeV α emitter as a result of ^{246}Cm bombardment with ^{13}C ions. In work begun in 1957 and concluded in 1958, a Russian group claimed to have produced an isotope of element 102 with an α decay energy of 8.8 ± 0.5 MeV by the bombardment of ^{241}Pu with ^{16}O.

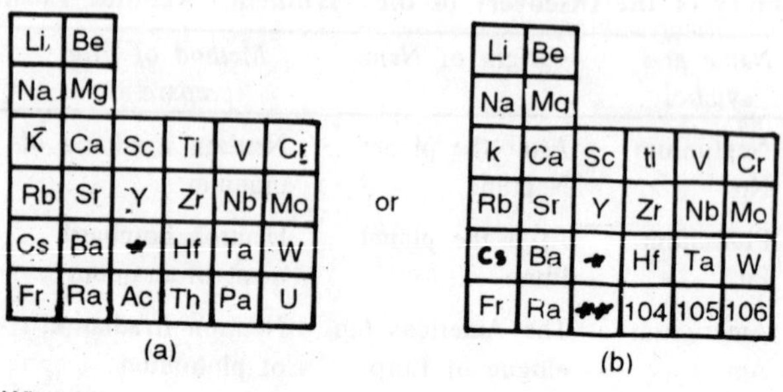

Where

* = Lanthanides =La, Ce, Nd, Pm, Sm, Gd, Ho, Er, Tm, Yb, Lu
** = Actinides =Ac. Th Pa, U, Pu, Cm, Bk, Cf, Fm, Md, No, Lw.

Fig. 9.4 Alternative positions of the actinides in the Periodic Table.

Evidence for the electronic configuration of these elements shown that the 5*f* atomic shell is filling in analogy to the lanthanides in which the 4*f* shell progressively fills. The electronic structure of the outermost shells of both lanthanides and actinides are shown together for comparative purposes in Table. 9.4. The trend for the early members of the series to exhibit higher valencies is attributed to the looser binding of the 5*f* shell compared with the 4*f* shell. The fact that the early members exhibit higher valencies is not entirely inconsistent with the lanthanides when it is remembered that the second and third elements in the lanthanide group, viz. cerium and praseodymium can exhibit a valency of four. It is assumed that , when discovered, elements 104, 105 and 106 will occupy the positions in the periodic table which were previously ascribed to thorium, protactinium and uranium.

(C) Oxidation States of the Actinides

(a) Trivalent

All of the actinides exhibit a trivalent state in aqueous solution with the exception of protactinium and thorium. U^{3+} is unstable and solutions of U^{3+} slowly evolve hydrogen. Np^{3+} compounds do not decompose water but are very easily oxidised by air to Np^{4+} . Pu^{3+} is stable to air but solutions are slowly oxidised as a consequence of their own radiation to form Pu^{4+}. Pu^{3+} is also readily oxidised by mild oxidising agents.

Table 9.4

Electronic Configuration of the Outer Shells of the Lanthanide and Actinide Elements

Lanthanides			*Actinides*		
Atomic number	*Name*	*Electronic configuration*	*Atomic number*	*Name*	*Electronic configuration*
57	Lanthanum	$5d$ $6s^2$	89	Actinium	$6d$ $7s^2$
58	Cerium	$4f$ $5d$ $6s^2$	90	Thorium	$6d^2$ $7s^2$
59	Praseodymium	$4f^3$ $6s^2$	91	Protactinium	$5f^2$ $6d$ $7s^2$
60	Neodymium	$4f^4$ $6s^2$	92	Uranium	$5f^3$ $6d$ $7s^2$
61	Promethium	$4f^5$ $6s^2$	93	Neptunium	$5f^4$ $6d$ $7s^2$
62	Samarium	$4f^6$ $6s^2$	94	Plutonium	$5f^6$ $7s^2$
63	Europium	$4f^7$ $6s^2$	95	Americium	$5f^7$ $7s^2$
64	Gadolinium	$4f^7$ $5d$ $6s^2$	96	Curium	$5f^7$ $6d$ $7s^2$
65	Terbium	$4f^9$ $6s^2$	97	Berkelium	$5f^8$ $6d$ $7s^2$ (or $5f^9$ $7s^2$)
66	Dysprosium	$4f^{10}$ $6s^2$	98	Californium	$5f^{10}$ $7s^2$
67	Holmium	$4f^{11}$ $6s^2$	99	Einsteinium	$5f^{11}$ $7s^2$
68	Erbium	$4f^{12}$ $6s^2$	100	Fermium	$5f^{12}$ $7s^2$
69	Thulium	$4f^{13}$ $6s^2$	101	Mendelevium	$5f^{13}$ $7s^2$
70	Ytterbium	$4f^{14}$ $6s^2$	102	Nobelium	$5f^{14}$ $7s^2$
71	Lutetium	$4f^{14}$ $5d$ $6s^2$	103	Lawrencium	$5f^{14}$ $6d$ $7s^2$

(b) Tetravalent

The only actinides with a reasonably stable valency of four are thorium which exhibits no other valency in solution, uranium and neptunium which are slowly oxidised by air, and berkelium. At first it seems strange to find a stable valency of four for berkelium where true rare earth characteristics should prevail. It is, however, not so surprising when it is remembered that its rare earth analogue, terbium, also exhibits a valency of four. Both of these elements have one electron in excess of a half full f shell. Pu^{4+} exists only in strong acid media, while Am^{4+} and Cm^{4+} exist only as the complex fluoride ion. Protactinium is partially reduced by zinc amalgam from its normal stable valency of five to an unstable tetravalent state which is very rapidly oxidised in solution by atmospheric oxygen.

(c) Pentavalent

The pentavalent state, Pa^{5+}, is the normal stable state of protactinium. Of the other actinides, neptunium has a complex ion NpO_2^+ which is stable at low and medium acidities but disproportionates at high acidities. The ions UO_2^+, PuO_2^+ and AmO_2^+ are all unstable and disproportionate.

(d) Hexavalent

The ions UO_2^{2+}, NpO_2^{2+} and PuO_2^{2+} are all stable. The UO_2^{2+} is difficult to reduce, but the other two are fairly easy to reduce. The ion AmO_2^{2+} is known, but it is reduced by its own radiation.

Colour of ions in solution. A notable feature of the actinides between uranium and americium, is the colour exhibited by the various oxidation states in aqueous solutions. For the more stable oxidation states these are:

Uranium is reddish-purple in the trivalent state, green in the tetravalent state and yellow-green in the hexavalent state.

Neptunium is purple in the trivalent state, green in the tetravalent state, blue green in the pentavalent state and light pink in the hexavalent state.

Plutonium is blue-violet in the trivalent state, purple in the pentavalent state and orange-brown in the hexavalent state.

Americium is pink in the trivalent state, yellow in the pentavalent state and amber-yellow in the hexavalent state.

As far as is known, the other actinides are colourless in solution.

(D) Formation of Complex Ions

The formation of ionic complexes with many common acid anions is important since it is the basis for many ion exchange and solvent extraction separation and purification processes.

In general, actinides in the trivalent state do not form neutral and anionic complexes with common mineral acid anions. The separation of trivalent actinide cations is usually accomplished by absorbing them on a cation exchange resin column and selectively eluting them with the ammonium salt of an α-hydroxy acid such as citric acid, lactic acid, or, better still, α-hydroxy-isobutyric acid. They may also be selectively eluted from a cation resin using a 20% ethyl alcohol solution saturated with HCI. Another useful method is solvent extraction using a chelating compound such as thenoyltrifluoroacetone (TTA) dissolved in a nonpolar organic solvent.

Actinides with valencies greater than three tend to form neutral and anionic complexes in solutions of nitric and hydrochloric acids. A notable exception is thorium which does not form neutral or anionic complexes in hydrochloric acid. Actinides exhibiting higher valencies also tend to form anionic complexes in dilute sulphuric acid ($\lesssim 0.1$ *M*). Neutral complexes can be selectively extracted by a number of organic solvents, notably tributyl phosphate, while anionic complexes can either be extracted using a solvent such as a high molecular weight amine or they can be absorbed on an anion exchange resin. Notable examples where some of these methods are used on a large scale are:

(a) Preparation of uranium concentrates from uranium ore:

The ore is leached with sulphuric acid and the solution passed through anion exchange columns from which the uranium is subsequently stripped with brine. Amine extraction of a sulphate leach is also used in some processes.

(b) Purification of uranium concentrates

Uranium is usually obtained in a pure state suitable for fuel element fabrication by dissolving the uranium concentrate in an excess of nitric acid, and solvent extracting the liquor with 20% to 30% TBP dissovled in a hydrocarbon diluent.

(c) Processing of spent reactor fuel elements to separate the fission products and recover uranium and plutonium

The fuel elements are dissolved in an excess of nitric acid and the solution extracted with approximately 30% TBP to extract both uranium and plutonium. The plutonium is recovered from the solvent by reducing the plutonium to the trivalent—non-complexing—state. The uranium is subsequently stripped from the solvent using very dilute nitric acid.

(d) Separation of fission products and recovery of thorium and fissile uranium-233 from irradiated thorium

Several methods have been proposed, many of them on a dual cycle system. After dissolution in nitric acid the solution can be first extracted with 5% TBP to recover the uranium, and then with 40% TBP to recover the thorium. Alternatively 40% TBP can be first used to extract both thorium and uranium, after which the thorium is selectively stripped with dilute nitric acid before the uranium is stripped using water.

10
Uses of Isotopes

In this Chapter some of the more important applications of isotopes will be outlined. These applications will be divided into three groups:

(1) Tracer applications where the nuclear properties of an isotope of an element are used as a label to study the behaviour of some material.

(2) Radiation applications where radioactive nuclides are employed as a source of radiation to either:

(*a*) produce some effect in a material, or
(*b*) make some measurement on a material.

(3) Dating techniques where the measurement of specific isotopes in a sample can be used to provide information about the age of the sample.

Tracer Applications

Isotopes may be used as labels in a variety of tracer applications both on a large industrial scale, involving the use of several curies at a time, or on a laboratory scale, requiring the use of only microcuries of material. While most applications involve the use of radioactive isotopes, this is not always the case. In some instances, where suitable radioactive isotopes are unavailable, separated stable isotopes, *e.g.* nitrogen-15 and oxygen-18 may also be used. In these latter applications assay is made using a mass spectrometer instead of by radiation measuring equipment.

Inherent advantages in the use of tracers are that they make possible measurements which would otherwise be difficult or impossible, and also the high degree of sensitivity which is possible using radiochemical methods of assays.

A great deal of the success in the use of tracers lies in the choice of a suitable nuclide. The more important factors which should be considered in the selection of a tracer for a particular application are:

(i) Chemical Properties. The tracer should be used in a form which is chemically compatible with the system to be studied and behave in a manner typical of the system under investigation.

(ii) Nature of Radiation. If radioactive, it must emit a radiation of a type and energy which can be readily measured under the conditions of the experiment, and unless necessary as a consequence of the conditions of the experiment, penetrating γ-radiation should be avoided where a β emitter is available. If two parameters are to be measured in the system using two tracers they should either emit different types of radiation (*cf.* Expt. 16), or similar radiations of different energy.

(iii) Half-life. The tracer must be chosen with a half-life long enough to permit the necessary measurements to be made, yet short enough so as not to create a lasting hazard.

(iv) Quantity of Activity. The amount of activity used should be the minimum consistent with the ability to make accurate measurements.

(A) Large Scale and Industrial Tracer Applications

(a) Siltation Measurements

The movement of mud and sand in river estuaries and harbours has been investigated in many countries using tracers. The procedure involved is to label a sample of the material with an insoluble form of the tracer in the laboratory, then to release it on the bed of the river or harbour, and to follow its movement under various tidal conditions with a detector suspended over the side of a boat. Because of the method of measurement, it is necessary that the tracer be a γ emitter, and since measurements may extend over a period of weeks, a moderate half-life is also necessary. Tracers which have been successfully used include scandium-46 ($t_{1/2}$ = 84 days), barium-140 ($t_{1/2}$ = 12.8 days), and chromium-51 ($t_{1/2}$ = 27.8 days). Quantities of activity involved are of the order of several curies.

(b) Movement of Underground Water

The direction and velocity of underground water flow can be traced by adding a radioisotope in the form of a water soluble compound to one bore hole and then taking samples at various time intervals from surrounding bores. Care must be taken in choosing an isotope which can not be precipitated from solution or absorbed by underground strata. The nuclide should preferably be a medium or high energy β emitter to simplify assay, and have a short half-life to avoid any lasting contamination. Iodine-131 with a half-life of eight days used as the iodide ion has been found generally suitable. Similar techniques have been used to measure porosity and flow rates in oil bearing formations, and also to measure flow patterns in surface waters *e.g.* in lakes and coastal regions.

(c) Leak Detection

The use of tracers to locate leaks in pipelines—especially water mains—is widely practiced. In the case of water pipes a water soluble, γ emitting nuclide with a short half-life such as sodium-24 ($t_{1/2}$ = 15 hours) is used. A plug of active solution is forced through the pipe under pressure, and the points where leakage into the surrounding soil has occurred are subsequently scanned from the surface using a γ detector (Fig. 10.1a). Where the pipe is too deep to permit this technique to be used the position of the leak can be found by emptying the fluid from the pipe section and then dragging a detector through. A refinement of this technique, which can be used on pipelines several miles in length, is the "go-devil" (Fig. 10.1b). This is a self-contained, battery operated detector and wire recorder which is pushed along the pipe by the fluid some distance behind a plug of radioactive solution. Cobalt-60 sources attached to the pipe at known intervals provide marker "blips" on the wire recorder as the go-devil passes. Where a leak exists and activity has been left in the soil, an additional blip is recorded and its position is defined relative to the cobalt markers.

(d) Measurement of Mixing Efficiency

Many industrial processes involve mixing two or more ingredients together, *e.g.* manufacture of paints, lubricating oils. etc. Incomplete mixing results in an unsatisfactory product while excessive mixing results in wasted time and plant facilities. The mixing efficiency using different impeller arrangement, and the rate at which homogeneous product can be obtained, is easily determined by inserting a small quantity of labelled

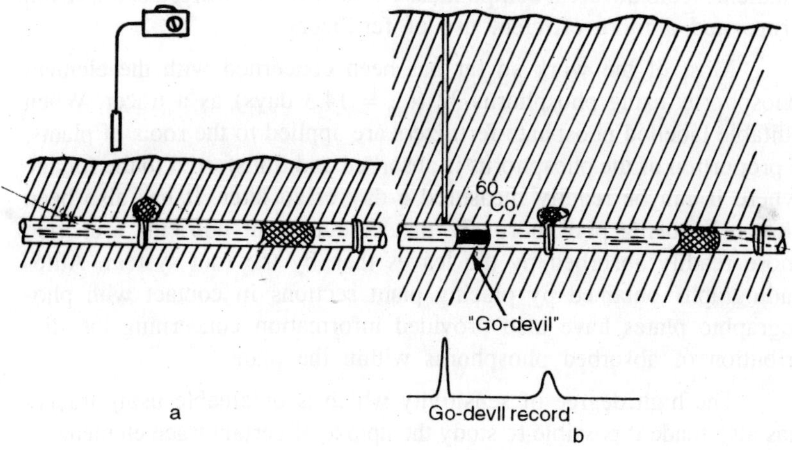

Fig. 10.1 Methods for the detection of leaks in underground pipes using radioactive tracers.

material and subsequently sampling the mix at various points over a period of time.

(e) Measurement of Wear

Tracers provide an extremely sensitive method whereby wear and erosion measurements may be made. An example is the inclusion of a radioactive tracer in a piston ring of an engine with subsequent radiochemical assay of samples of oil drawn from the sump.

(f) Interface Marking

The interface between two different oil fractions pumped through the same pipeline can be marked by the addition of an organic compound containing a short lived γ emitter, e.g. bromine-82 ($t_{1/2}$ = 35.3 hours). The arrival of the labelled oil—and hence the junction of the two fractions—is recorded by a detector on the outside of the pipe. During storage the tracer decays to negligible levels of activity. Small solid semibuoyant balls containing a long-lived γ emitter have also been used. These are subsequently recovered for re-use.

(B) Use of Tracers in Agriculture

A large amount of research has been carried out in recent years on the uptake rates and the mechanisms whereby plants absorb essential elements from different compounds. These have, in turn, led to improved crops and a more efficient use of fertilizers.

Most of the work so far has been concerned with the element phosphorus using phosphorus-32 ($t_{1/2}$ = 14.3 days) as a tracer. When suitable labelled phosphate fertilizers are applied to the roots of plants, a proportion of the phosphorus is absorbed and transferred to the foliage where it can be readily measured a day or so later. It has also been shown by using phosphorus-32 that certain plant nutrients are often more readily absorbed by the leaves than by the root system. Autoradiographs obtained by placing plant sections in contact with photographic plates have also provided information concerning the distribution of absorbed phosphorus within the plant.

The high degree of sensitivity which is obtainable using tracers has also made it possible to study the uptake of certain trace elements—or "micronutrients"—such as cobalt, copper, manganese, molybdenum and zinc. Using tritium—a radio-active isotope of hydrogen—it has been possible to measure the rate of movement and distribution of water taken up by plant roots. Plant respiration has also been investigated using carbon-14 labelled carbon dioxide.

(C) Use of Tracers in Biology and Medicine

A number of biological processes in both plants and animals have been elucidated using tracer techniques. Notable among these has been the use of carbon-14 to determine the steps involved in photosynthesis. In the medical field traces have not only provided a method for the study of many physical and biochemical processes, but have also provided valuable diagnostic techniques.

In biological applications complex organic compounds are frequently required which cannot be conveniently synthesized in a chemical laboratory. For such applications a process known as labelling by biosynthesis is employed. In this process some form of life which produces the compound concerned is "fed" with an isotope which it is wished to incorporate in the compound. A few examples of some labelled compounds which have been prepared in this manner are: haemoglobin labelled with active iron form the red blood cells of animals fed with active iron, casein labelled with phosphorus-32 from milk obtained from a cow into which an active inorganic phosphate had been injected, and penicillin labelled with sulphur-35 prepared by adding sulphate containing ^{32}S to the medium in which the penicillin in grown.

Examples of tracers used in the medical field are:

Iodine-131 ($t_{1/2}$ = 8 days). Iodine-131 is frequently used to study the functioning of the thyroid gland which is located in the neck, and which manufactures an iodine compound called thyroxine. The uptake of iodine by the thyroid can be determined by making measurements with a scintillation counter placed close to the patients neck over a period of 48 hours after the oral administration of five microcuries of iodine. It is also possible, by using a collimated detector, to plot the shape and size of the thyroid following a dose of iodine-131.

Iodine-131 as labelled albumin has also been used to locate brain tumours which preferentially absorb the iodinated albumin. By scanning the skull with a collimated detector information concerning the existence, size and position of a tumour can be obtained.

Phosphorus-32 ($t_{1/2}$ = 14.3 days) and *arsenic*-74 ($t_{1/2}$ = 18 days). Both of these isotopes have also been applied to the identification and location of malignant tumours. Arsenic-74 is particularly interesting since it is a position emitter. Two collimated scintillation detectors connected in coincidence and located on opposite sides of the patient can be used to detect the position annihilation γ-rays, thereby locating tumours more accurately.

Iron-59 ($t_{1/2}$=45 days) can be used to study the production of red corpuscles in bone marrow and to determine the average life of these corpuscles in the body.

Cobalt-58 ($t_{1/2}$ = 71 days) has been used to study the function of vitamin B_{12} in the body and is used diagnostically in the detection of pernicious anaemia, which results from the inability of the body to absorb vitamin B_{12}. Cobalt-58 labelled vitamin B_{12} has also been used to investigate preparations for promoting the absorption of this vitamin.

(D) Use of Tracers in Chemistry

Tracers, both stable and radioactive, have been incorporated in many compound to study chemical reactions. Both inorganic and organic reactions have been studied in detail using tracer techniques. In the case of organic reactions, where molecules contain several atoms of the same element in different positions, the fate of any one of these atoms in the course of a chemical reaction can be followed by incorporating an isotopic tracer in the relevant position. Isotopes used to study organic reactions are radioactive carbon-14 ($t_{1/2}$ = 5730 years) and the inactive nuclides : oxygen -18, nitrogen-15 and deuterium. A mass spectrometer must be used to detect and measure these stable tracers. Radioactive tritium can be used as a tracer for hydrogen, but because of its relatively large mass difference compared with hydrogen, it often exhibits reaction rates differing from those of hydrogen, and results obtained using tritium are therefore often less valid. Labelling of carbon compounds seldom requires total synthesis as there are several hundred labelled carbon-14 compounds commercially available.

Typical of some chemical applications of tracers are:

(a) Study of Exchange Reactions

Here atoms of an element exchange places with other atoms of the same element in a different chemical species. Studied of these reactions are only possible by using different isotopic species of the element concerned. A simple example is the use of deuterium to establish the exchange of hydrogen atoms between hydrogen gas and water in the presence of a catalyst

$$H_2O + D_2 \xrightarrow{Pt} HDO + HD$$

Iodide ion has also been shown to exchange with the iodine in alkyliodides dissolved in ethanol

$$C_2H_5I _ {}^{131}I^- \rightarrow C_2H_5\ {}^{131}I + I^-$$

The nonequivalence of the sulphur atoms in $S_2O_3^{2-}$ has been illustrated by the fact that the two sulphur atoms do not exchange with one another, and that only the one which can be precipitated by acid exchanges with the sulphur in HS^-.

Another type of exchange reaction is the electron transfer reaction which involves the exchange of oxidation states between two isotopic species:

$$Ce^{3+} + {}^{144}Ce^{4+} \rightarrow Ce^{4+} + {}^{144}Ce^{3+}$$

This can be measured by mixing one isotope in one oxidation state with another isotope of the same element in another oxidation state, then separating the two oxidation states after a time interval and measuring their isotopic composition. This has produced some unexpected results. For example, while the exchange between Cu^+ and Cu^{2+} is too rapid to measure, the exchange between Ce^{3+} and Ce^{4+} takes many minutes.

(b) Investigation of Reaction Mechanisms

A large number of reactions have been studied by labelling part of a molecule with a tracer prior to a reaction, then seeking the tracer in the reaction products. A sample example is the hydrolysis of an ester. In this reaction two mechanisms are possible:

$$\text{RCOO}\left[\text{R}^1 + \text{HO}\right]\text{H} \rightarrow \text{RCOO}\left[\text{H} + \text{HO}\right]\text{R}^1$$
$$\text{RCOO}\left[\text{R}^1 + \text{H}\right]\text{OH} \rightarrow \text{RCO}\left[\text{OH} + \text{H}\right]\text{OR}^1$$

In the hydrolysis of amyl acetate in oxygen-18 labelled water, all of the oxygen was found in the acid which indicated that the reaction proceeds exclusively by the process illustrated in the second of the equations shown above.

Possibly the most spectacular use of tracers to study a reaction mechanism, is the use of carbon-14 labelled carbon dioxide to determine the various complex steps in the phenomenon of photosynthesis. Algae were fed on labelled carbon dioxide and exposed to light for various intervals of time after which the reaction was stopped and the products separated by paper chromatography. An autoradiograph, obtained by placing the chromatogram in contact with a photographic plate, was used to locate the various radioactive products, which progressively appeared as the exposure of the system to light was increased. From experiments such as this the complete carbon cycle in photosynthesis has been built up.

(c) Properties of Metal Ions

Tracer techniques have been widely used to study the chemistry of large number of ionic species. These techniques have been particularly useful in the development of solvent extraction and ion exchange processes for the commercial recovery of some of the rarer metals. The advantages of using tracer techniques in these investigations lie in the speed and accuracy which is possible by comparison with more conventional techniques. Tracer techniques are also very useful in the study of many fundamental chemical properties. For example, many metal ions form complexes in acid solution. In the case of thorium in nitric acid solution complexes of the form $(Th\ (NO_3)_n)^{(4-n)+}$ will be formed. Information concerning the nature of these complexes at different acid concentrations can be determined by two techniques:

(i) Electromigration experiments carried out by soaking a filter paper strip in nitric acid of the concentration under investigation, placing a small quantity of thorium-234 tracer at the centre, and applying a potential between the ends of the strip. After a period of time the potential is removed, the strip dried and scanned with a counter to measure the direction (cationic or anionic) and distance the complexes have moved.

(ii) Solvent extraction procedures can also be used to provide similar information by studying extractability into solvents which selectively extract withe cationic, neutral or anionic complexes. It is also possible using an extension of solvent extraction techniques to determine the number of acid ligands in the complexes and the formation constants of the complexes.

(d) Radiometric Titrations

Tracers may be used to determine the end point in a titration where the reaction produces a compound which is either insoluble or extractable into a solvent phase. If, fro example, silver is to be determined by titration with standard hydrochloric acid, silver-110 tracer is added and the specific activity of the solution is recorded on a graph of activity vs. hydrochloric acid. As the titration proceeds silver is removed from the solution by precipitation, so that if the specific activity of the solution is measured after the addition of known volumes of hydrochloric acid, a linear relation between residual activity and hydrochloric acid will be obtained. Extrapolation to zero activity corresponds to the end point of the titration.

(e) Isotope Dilution Analysis

This is an analytical method which can be employed to determine the total quantity of some element or compound which it is either difficult or impossible to isolate in quantitative yield. The technique involves isolating and estimating only a fraction of the material to be assayed and to use a tracer to determine what this fraction is. This is done by adding a known amount of a tracer which is either ran isotope of the element to be assayed, or a labelled compound of the compound to be assayed, before the isolation begins, and then measuring the amount of tracer in the final product. For example, carbon-14 labelled naphthalene can be added to coal tar residues prior to the estimation of its naphthalene content. A variation of this technique which is applied to the estimation of very small amounts of specific isotopes *e.g.* in geochronology consists of adding a known amount of a different isotope of the same element and measuring the isotope ratios after isolation of part of the element.

Although it is not, strictly speaking, and "isotope dilution" method, and analogous approach has been used to determine fluid volumes in closed systems, such as, for example, the measurement of the total blood volume of a human. This is one case where the direct approach to the problem would have unfortunate side effects. Total blood volume can, however, be measured by injecting a known activity of sodium-24 ($t_{1/2}$ = 15 hours) into the blood stream in a saline solution. After allowing sufficient time for it to be homogeneously mixed with the blood, a small measured volume of blood is removed and the sodium-24 activity is measured.

Radiation Applications

(A) Use of Radiation to Produce Some Effects in a Material

We have already seen in Chapter 7 radiation can produce many chemical effects in materials through which they pass. At the present time the use of radiation to promote chemical reactions (*e.g.* polymerisation) is more of academic rather than commercial interest. The effects which radiation can produce in living materials and organisms is of considerable importance. Among these are:

(a) Sterilisation

The use of radiation to sterilise a number of products is rapidly growing. About one million rads are necessary to destroy bacteria, and about one tenth of this amount is necessary to kill insects and their eggs in grain. Radiation sterilisation finds its greatest commercial application in the field of pharmaceuticals. This form of sterilisation is virtually the only one possible for heat sensitive compounds such as antibiotics. There are also many advantages in being able to package drugs and surgical instruments under non-sterile conditions and to then subsequently completely sterilise the contents of the sealed packages without any possibility of subsequent recontamination. Partial sterilisation of food to enhance its keeping qualities is also at present under economic consideration.

(b) Radiotherapy

Living cells can be destroyed by radiation. Malignant cells are generally more susceptible than normal cells and it is partly for this reason that some degree of success has been achieved in the treatment of cancer using radiation—in particular that from cobalt-60 sources. After a cancer has been located—and here tracers can provide and important diagnostic role the patient is placed beneath a strong (~ 1 kilocurie) collimated cobalt-60 source on a table which is in continuous motion such that the malignant area receives a continuous dose, while the surrounding area receives a lesser one. In another method of treatment small encapsulated sources are implanted in the tumour tissue. Because they are in the immediate proximity to the area to be irradiated, and because they provide a more or less continuous irradiation, sources of only a few millicuries are generally required. It is also customary in these cases to use a much less penetrating source of γ radiation. Gold-198 (0.41 MeV) is a common choice.

(c) Radiation Induced Mutations

Mutations can be induced in plant species as a result of radiation. Under normal circumstances mutations occur without any irradiation. Most are detrimental while a few lead to improved strains. Selective breeding of these can lead to new and improved species. Irradiation of plants or their seeds has the effect of greatly accelerating the mutation rate and thereby making possible the development of improved strains in a fraction of the time which would otherwise have been necessary.

Some applications where radiation sources are used to produce an effect in inanimate matter include:

(a) Elimination of Static Electricity

Small β emitting sources are frequently used to remove static changes which build up in weaving mills. The source, which is hung over the loom, ionises the air in its vicinity and thus allows static charges to leak away. Static eliminators are also used as a safety measure in the extrusion of cordite.

(b) Luminous materials and light sources

The use of luminous radioactive paint on watch and clock dials is familiar. Not so well known are the signalling lamps which have been developed using the fission product inert gas krypton-85. These can produce light visible over a distance of half of a mile and are suitable where a beacon is required to produce a continuous maintenance free light source in some remote spot where electrical energy is not available. Radioactive tritium is also incorporated in plastic to produce luminous light switch fascias and in other semi-ornamental applications.

(c) Nuclear Batteries

A method for the direct conversion of radioactive energy into electrical energy has been sought after for many years and several proposals have been investigated. The most obvious and direct arrangement is the use of two electrodes—one coated with a β emitter—in an evacuated chamber. The β "current" inside the chamber results in an equivalent electron flow in an external circuit. Such devices are capable of producing very large potentials but at infinitesimally small currents and are therefore of little practical value. Two types of cell based on semiconductor p-n junctions however do provide a basis for practical application.

In one of these, a pure β emitter such as promethium-147 is used to promote a flow of electrons across the junction. Some 10^5 electrons per β-particle are produced and an effective potential of 0.2 volts can be obtained. A limitation of the cell is the radiation damage produced by the β-particles in the semiconductor. In another similar type of cell a radioactive light source is used in conjunction with a photosensitive p-n junction semiconductor.

It is improbable that nuclear batteries will ever replace the more conventional dry cells in everyday applications. Their potential value lies in special applications where small size and long term service without replacement is necessary. One such application is as a power source in space satellites.

(B) Use of Radiation to Make Some Measurement

The attenuation of β- and γ-radiation is the basis for a number of applications. These include:

(a) γRadiography

A γ emitting source placed on one side of an object will form an image on a piece of film placed on the other side corresponding to the relative attenuation experienced by the γ-rays as they passed through the object. The technique is very similar to the more familiar X-ray methods, but has the advantages of greater portability and lower capital cost. Isotopes emitting γ-rays of different energy are used for different applications. Soft γ-radiation from thulium-170 is suitable for dental "X-rays" and for radiography of aluminium, while iridium-92 is suitable for steel of from 0.5 cm to 8 cm. For steel up to 15 cm in thickness cobalt-60 is suitable. The main use of these sources is for the inspection of welds in large engineering projects and in places of limited accessibility.

(b) Thickness Gauging

The attenuation experienced by radiation is a function of the thickness of the absorbing medium through which it passes. It is therefore possible to use a measure of the attenuation as a means to determine the thickness of an absorbing medium.

ssion thickness gauges may employ either a β source (paper, linoleum and metal foils) or a γ source (heavy gauge sheet metal). The source is placed on one side of the material to be measured and the detector on the opposite side (Fig. 10.2a). For optimum results a source of radiation should be used which has an absorption half-thickness in the material to be measured which is approximately equal to one-third of the thickness to be measured. Not only do transmission gauges represent an ideal method for the continuous monitoring of a process, but they may be used to actuate a servo mechanism to control any deviation in the thickness of the product.

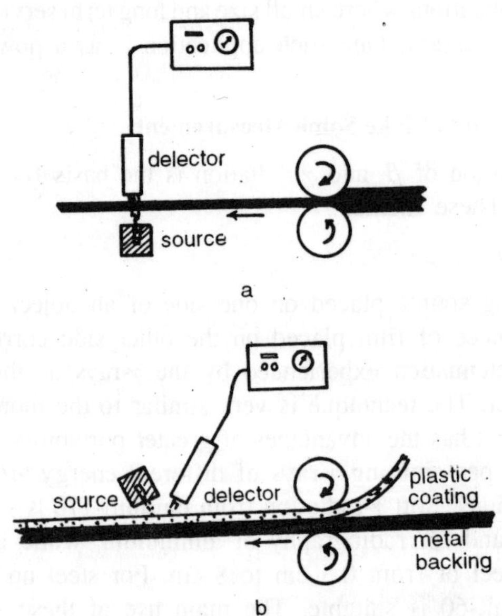

Fig. 10.2. Measurement of thickness.
(a) Transmission gauge; (b) backscatter gauge.

Another type of thickness gauge is the backscatter gauge in which both source and detector are used on the same side of the material to be measured. These are used much less frequently. However, they do make possible the measurement of the thickness of coatings on a backing material of higher atomic number (Fig. 10.2b).

(c) Density Measurement

A source and detector on opposite sides of a tube can be used to monitor the density of the material flowing through the tube, since the attenuation over a fixed distance is a function of the density of the medium. Backscatter gauges have also been developed for measuring the density of rock strata in oil wells.

(d) Measurement of Liquid Levels

The level of liquids in closed containers such as those at oil refineries are often measured by the use of a source and detector sliding on vertical rails on opposite sides of the container. As the source and detector are moved upward past the surface a sudden increase in count rate occurs. Similar techniques have also been employed to monitor the contents of packets.

Dating Techniques

Measurement of the age of samples form the past on an absolute time scale has remained a challenge for many centuries, and it is perhaps a little ironical that the solution to problems such as the age of the earth itself should lie in the study of atoms. The first attempts to estimate geological ages were made in 1907, when Boltwood estimated the age of minerals containing uranium by measuring their lead content on the assumption that it had all originated from the radioactive decay of uranium. Since that time, a better understanding of radioactive processes together with greatly improved analytical techniques—such as isotope dilution analysis, mass spectrometry and sophisticated electronic counting devices have led to improved accuracy and greatly extended applications.

Dating methods which have been developed and which will be discussed here are:

(a) Geochronology—the determination of the age of minerals and geological structures which form the earth's crust (measured in millions of years).

(b) Carbon dating—the determination of carboniferous material, e.g. archeological samples (of use up to tens of thousands of years).

(c) Tritium dating—the determination of the age of natural water (up to tens of years).

(A) Geochronology

The principle upon which the age determination of geological samples is based is simple. A naturally occurring radioactive nuclide decays to a daughter, the quantity of which increases with time. From a measurement of the ratio of the daughter to th parent, together with a knowledge of the half-life of the parent, it is possible to estimate the age of the system.

If **P** is the number of parent atoms left, and **D** is the number of stable daughter atoms formed, then the original number of atoms must have been **P** + **D**. Substituting these values in the radioactive decay formula:

$$\log_{10}\left(\frac{N_t}{N_0}\right) = -0.4343\,\lambda t$$

we have

$$\log_{10}\left(\frac{\mathbf{P}}{\mathbf{P}+\mathbf{D}}\right) = -0.4343\,\lambda t$$

or

$$t = \frac{2.303}{\lambda}\log_{10}\left(1+\frac{\mathbf{D}}{\mathbf{P}}\right)^{*}$$

The method is restricted to rock structures which:

(i) contain some radioactive parent nuclide,

(ii) constitute a chemically sealed system in which there has been no loss or gain of parent, daughter, or any intermediate nuclide in a decay chain,

(iii) initially contained none of the daughter, or if some daughter was present, that it is possible to calculate what this amount was.

Some of the systems studied are:

(a) Uranium-lead

$$\left\{\begin{array}{l} {}^{238}\text{U},\ t_{1/2} = 4.51\cdot 10^{9}\ \text{yr},\ \lambda = 1.54\cdot 10^{-10}\ \text{yr}^{-1} \\ {}^{238}\text{U},\ t_{1/2} = 7.1\cdot 10^{8}\ \text{yr},\ \lambda = 9.71\cdot 10^{-10}\ \text{yr}^{-1} \end{array}\right\}$$

Both the ^{238}U → intermediates → ^{206}Pb, and the ^{235}U → intermediates → ^{207}Pb systems can be employed in uranium bearing minerals and have led to consistent results.

---

* This formula applies to all but the potassium-argon system which involves branched decay.

(b) Thorium–lead

$$\{ {}^{232}\text{Th},\ t_{1/2} = 1.39 \cdot 10^{10}\,\text{yr},\ \lambda = 4.99 \cdot 10^{-11}\,\text{yr}^{-1} \}$$

The decay of ^{232}Th to its ultimate product ^{208}Pb can also be used. In cases where both the uranium and thorium methods have been applied to the same sample (*e.g.* uraninites) consistent results have usually been obtained. A complication of this system is the presence of nonradiogenic ^{208}Pb which it is believed usually amounts to 10% of the total ^{208}Pb present.

(c) Uranium–helium, thorium–helium

Because the emission of helium nuclei (α-particles) is one of the modes of decay, a determination of the amount of helium in a uranium or thorium mineral provides an alternative approach. Early results using volumetric techniques were not encouraging but more recent determinations using isotope dilution analysis have provided more consistent results. In general, however, the uranium-lead and thorium-lead methods are considered preferable.

(d) Potassium–argon

There is complication in this method due to the dual decay mechanism of potassium-40 (Fig. 10.3).

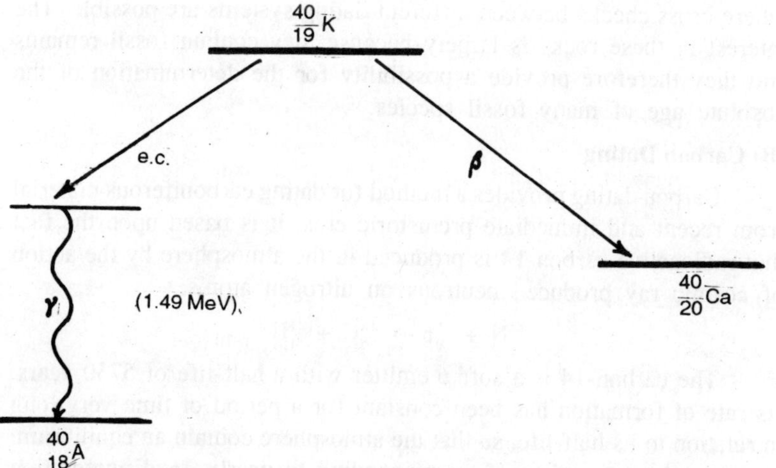

Fig. 10.3. Decay mechanism of potassium-40.

The formula relating time to the argon and potassium isotope ratios is:

$$t = \frac{2.303}{\lambda_{e.e.} + \lambda_{\beta}} \log_{10}\left(1 + \frac{\lambda_{e.e.} + \lambda_{\beta}}{\lambda_{e.e.}} \cdot \frac{^{40}A}{^{40}K}\right)$$

and the partial decay constants for electron capture and β decay are taken as:

$$\lambda_{e.e.} = 5.85 \cdot 10^{-11}\ yr^{-1}$$
$$\lambda_{\beta} = 4.72 \cdot 10^{-10}\ yr^{-1}$$

In many instances both uranium and potassium occur in the same materials and a favourable comparison of ages using the tow methods has been possible.

(e) Rubidium-strontium

$$(t_{1/2} = 4.85 \cdot 10^{10}\ yr,\ \lambda = 1.43 \cdot 10^{-11} yr^{-1})$$

In the system $^{87}Rb \rightarrow {}^{87}Sr$, allowance must be made for nonradiogenic ^{87}Sr. This is readily accomplished by a measurement of ^{88}Sr present. Good agreement between this system and the potassium-argon system in micas, has provided a cross check to prove the validity of the system.

A good deal of interest has recently been centered upon the age determination of sedimentary rocks. The dating of these rocks is, in general, less reliable than for igneous rocks due to their less impervious nature, and to the fact that it is much more difficult to find samples where cross checks between different dating systems are possible. The interest in these rocks is largely because they contain fossil remains and they therefore provide a possibility for the determination of the absolute age of many fossil species.

(B) Carbon Dating

Carbon dating provides a method for dating carboniferous material from recent and immediate prehistoric eras. It is based upon the fact that radioactive carbon-14 is produced in the atmosphere by the action of cosmic ray produced neutrons on nitrogen atoms:

$$^{14}_{7}N + {}^{1}_{0}n \rightarrow {}^{14}_{6}C + {}^{1}_{1}H$$

The carbon-14 is a soft β emitter with a half-life of 5730 years. Its rate of formation has been constant for a period of time very long in relation to its half-life, so that the atmosphere contain an equilibrium concentration of carbon-14 corresponding to nearly 16 disintegration per minute per gram of carbon. All living plants and animals contain this equilibrium concentration, but when a plant or animal dies no further exchange between it and its environment can occur and the carbon-14 content decreases as a result of radioactive decay. After 5730 years there will remain only about 8 dis/min/g, and after another half-life only about 4 dis/min/g will remain and so on. A measure of the specific activity of a carbon sample therefore provides a measure of the age of the sample.

The determination of the activity is complicated by both the low activities which have to be measured and by the low energy of the β-particles which are emitted. Methods which have been used include counting an elemental carbon deposit which suffers the disadvantage of low counting efficiency, liquid scintillation methods which involve an elaborate chemical synthesis, and the counting of gaseous sample in a proportional counter. Over 95% of all dates determined at present use this latter method. The gas used to fill the proportional counter acts as both sample and counting medium. Carbcn dioxide obtained by combustion of the sample and methane obtained by combustion of the sample and methane obtained by reacting carbon dioxide with hydrogen gas in the presence of a catalyst are the gases most commonly employed. The extra step involved in the use of methane is offset by the better counting characteristics obtained with this gas. A knowledge of the counter volume and a measurement of the gas pressure defines the total carbon content of the sample, while the count rate defines the amount of carbon 14 present.

Since very low count rates are involved long counting periods (~ 24 hours) are necessary. The background count is reduced by the use of both heavy shielding and by a rig of Geiger tubes arranged in anticoincidence with the proportional counter.

(C) Tritium Dating

Tritium is a radioactive isotope of hydrogen with a half-life of 12.26 years. Like carbon-14, it is produced in the atmosphere by the action of cosmic rays although the mechanism for its production is not certain. One suggestion is that it is formed by the interaction of fast neutrons with nitrogen:

$$^{14}_{7}N + ^{1}_{0}n \rightarrow ^{12}_{6}C + ^{3}_{1}H$$

The proportion of tritium in ordinary water is very small (~ 1 part in 10^{18}) and before it can be measured an isotope enrichment must first be performed. Applications of tritium dating appear to be very limited. One useful region of research is in the investigation of underground water supplies to determine whether they originate from recent rainfalls, or whether they come from large underground reserves.

11

Experimental Nuclear Chemistry

Radiochemical Hazards

The ionisation of atoms and molecules results from the passage of nuclear radiation through matter, a single α-, β- or γ-ray producing about 100,000 ion pairs. A secondary effect of ionisation in molecules, is the rupture of chemical bonds causing the destruction of the chemical compounds from which living tissue is composed.

In handling radioactive materials, two distinct types of hazard should be distinguished and appreciated. These are:

(a) Contamination hazards involving ingestion of very small quantities of radioactive material, and

(b) Radiation hazards involving exposure to some external source of radiation.

Of the three types of radiation with which we are concerned, α-particles have the shortest range—though by no means the lowest initial energy—and dissipate their energy in an extremely short distance, producing intense ionisation. The range is only a few centimeters in air and a fraction of a millimeter in tissue. They thus represent a serious contamination hazard, but there is virtually no radiation hazard associated with them. The considerably longer range of β-particles in air and their range of a few millimeters in tissue, renders them not only a serious contamination hazard but also a radiation hazard—particularly as far as exposed organs e.g. the eyes, are concerned. The much greater range and penetration of γ-rays on the other hand, renders them primarily a radiation hazard.

(A) Contamination Hazards

The factors which influence the radio-toxicity of a nuclide as a contamination hazard if ingested orally, by inhalation, by absorption through the skin, or directly into the blood stream—as a result of an accident—are:

(a) The type of radiation emitted, α and β being the worst.

(b) The physical properties of the material, for example the chances that a compound ingested orally will enter the blood stream, and so be deposited in a vital organ, will be reduced if it is an insoluble compound.

(c) The chemical properties of the nuclide. Because of their varying chemical properties, isotopes of different elements may be deposited in different parts of the body or not retained at all. Elements in the same periodic group as calcium will, for example, accumulate in the bone where they will irradiate the blood forming cells in the bone marrow. It is largely for this reason that strontium-90 and radium-226 come very high in order of toxicity.

(d) Half-life of the nuclide. Besides the radioactive half-life we also have to consider the *biological half-life* which is defined as the time required for half of the material initially absorbed by the body to be expelled. A composite "effective" half-life which takes into account both radioactive and biological half-lives can be determined from:

$$\frac{1}{\text{effective half-life}} = \frac{1}{\text{radioactive half-life}} + \frac{1}{\text{biological half-life}}$$

A few examples of effective half-lifes are given in Table 11.1. Because of biological retention properties associated with many nuclei there is a long term accumulative effect and, therefore, the quantities of these materials which constitute a hazard are often extremely low.

Table 11.1

Nuclide	*Concentrates*	*Half-life (in days)*		
		Radioactive	*Biological*	*Effective*
Sodium-24	Whole body	0.62	19	0.61
Iodine-131	Thyroid	8.1	180	8.1
Strontium-90	Bone	$9.1 \cdot 10^3$	$3.9 \cdot 10^3$	$2.7 \cdot 10^3$
Radium-226	Bone	$5.9 \cdot 10^5$	$1.6 \cdot 10^4$	$1.6 \cdot 10^4$

In considering the various contributing factors mentioned above, nuclides may be roughly divided into group according to their degree of radiotoxicity. The common ones are classified in this way together with a rough guide to the working limits for normal chemical work (Table 11.2). In most countries the amount of active material which can be handled under various laboratory conditions are a matter of legislation and vary to some degree. The suggested limits in Table 11.2 will, in many cases, be on the conservative side. The term "ordinary laboratory" is intended to imply the type of senior teaching or research laboratory found in most universities. The term "special laboratory" is intended to cover a laboratory fitted with a non-porous floor covering, smooth washable walls, laminated plastic covered bench tops, forced ventilation—resulting from adequate fume hood facilities, a shielded safe for radioactive source storage, radioactive waste disposal arrangements and available radiation monitors. With additional glove box or shielded cell facilities considerably higher levels than those suggested can be safely handled. The amounts also depend very much upon the type of operation being carried out and upon the skill and experience of the person concerned.

Table 11.2

Radio toxicity	*Nuclide*	*Recommended limits*	
		Ordinary laboratory	*Special laboratory*
High	^{90}Sr, ^{210}Ph, ^{210}Po, ^{226}Ra, ^{227}Ac, ^{233}U, ^{235}U, ^{239}Pu, ^{241}Am, ^{242}Cm	10 μc	1 mc
Medium	^{45}Ca, ^{59}Fe, ^{89}Sr, ^{91}Y, ^{106}Ru, ^{131}I, ^{140}Ba, ^{144}Ce, ^{234}Th, (natural) Th and U	100 μc	10 mc
Low	^{3}H, ^{7}Be, ^{14}C, ^{24}Na, ^{32}P, ^{35}S, ^{36}Cl, ^{42}K, ^{46}Sc, ^{48}V, ^{51}Cr, ^{55}Fe, ^{56}Mn, ^{60}Co, ^{59}Ni, ^{64}Cu, ^{65}Zn, ^{71}Ge, ^{76}As, ^{86}Rb, ^{95}Zr, ^{96}Tc, ^{99}Mo, ^{103}Pd, ^{105}Ag, ^{109}Cd, ^{111}Ag, ^{113}Sn, ^{127}Te, ^{129}Te, ^{137}Cs, ^{140}La, ^{182}Ta, ^{181}W, ^{185}W, ^{191}Pt, ^{193}Pt, ^{196}Au, ^{198}Au, ^{200}Tl, ^{201}Tl, ^{204}Tl, ^{203}Pb	1 mc	100 mc

(B) Radiation Hazards

The radiation hazards associated with most chemical and tracer work will be small or negligible. They mainly arise from isotope stock storage and from sealed sources such as those used as radiation sources for radiation chemistry studies. Under no circumstances should the combined effect of all the sources of activity be permitted to involve a weekly gross exposure to any person in excess of 100 milliroentgens and, indeed, the aim should be to maintain conditions such that the total radiation dose cannot exceed a tenth of this value. Film badges can—and frequently are—worn in order to measure the accumulated weekly dose. They do, however, only measure what has already occurred, and in this respect, they are a poor substitute for careful radiation monitoring to anticipate radiation exposure.

Since the range of α-particles in air is only a few centimeters shielding from them is unnecessary. However, one must appreciate that most α emitters also have γ-radiation associated with them. In the case of β-particles, which have a finite range, one centimeter of "perspex" is adequate shielding and is, in fact, preferable to the use of materials of high atomic number which promote the production of bremsstrahlung. In the case of strong β, γ source an inner shield of "perspex" or aluminium is preferable to lead alone. Where high levels of γ-radiation exist, shielding in the form of lead bricks or concrete blocks should be used. The thickness of lead required to attenuate γ-radiation can be gauged from Fig. 11.1.

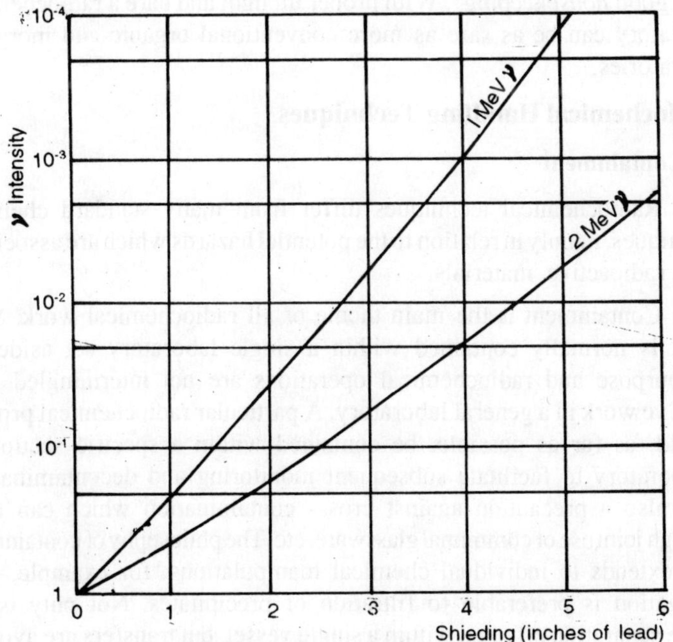

Fig. 11.1. Attenuation of γ-radiation by lead.

Five inches of concrete can be taken as equivalent to one inch of lead. As an indication of the order of magnitude of the γ-radiation levels associated with α, γ and β, γ sources it can be assumed that one curie of most α, γ and β, γ emitters will give a radiation level of between 0.3 and 1.3 roentgens per hour at one meter. The actual value will depend upon the proportion of decay transitions which involve γ emission and upon the energy of the emitted γ-ray.

Exposure to radiation hazards is minimised by one or more of the following methods:

(a) Using minimum necessary quantities of active material and limiting exposure time.

(b) Making full use of the inverse square law (Expt. 6) by remaining at a distance and by using forceps or tongs where necessary to achieve this end.

(c) Employing adequate' shielding—particularly in the case of fixed irradiation installations.

Rules and regulations concerning the use of radioactive material can at no time substitute for anticipation of hazards, care in manipulations and "good housekeeping". With proper thought and care a radiochemical laboratory can be as safe as more conventional organic and inorganic laboratories.

Radiochemical Handling Techniques

(A) Containment

Radiochemical techniques differ from many standard chemical techniques, mainly in relation to the potential hazards which are associated with radioactive materials.

Containment is the main theme of all radiochemical work. Such work is normally contained within a single laboratory set aside for the purpose and radiochemical operations are not intermingled with inactive work in a general laboratory. A particular radiochemical process should, as far as possible, be contained within a specific section of a laboratory to facilitate subsequent monitoring and decontamination. It is also a precaution against cross- contamination which can arise through joint use of communal glassware, etc. The philosophy of containment also extends to individual chemical manipulations; for example, centrifugation is preferable to filtration of precipitates. Not only is the phase separation effected within a single vessel, but transfers are avoided and the contaminated pieces of glassware are kept to a minimum. Liquid transfers—except those of low specific activity and large volume—are effected by transfer pipettes with rubber bulbs rather than by pouring, so that the liquid is at all times contained. The higher the level of activity handled the greater is the emphasis on containment. Above the millicurie level chemical operations are usually performed within the confines of a fume hood and above the curie level they are carried out in a totally enclosed, air tight "glove box" or—where there is a high γ activity—in a shielded "cell".

(B) Secondary Precaution

It is customary in radiochemical work, to provide some form of second line of defence against a contamination hazard resulting from an accidental spill or minor mishap. Common sources of minor contamination are the "splitting" of a solution during evaporation, a drop of solution falling from the end of a transfer pipette, or simply a contamination hazard which may arise through a piece of used equipment *e.g.* pipette, glass stirring rod, tongs *etc.*. being laid down on the bench. With small scale tracer work it is usual to place a piece of thin polyethylene sheet on the bench in the immediate working area, so that at the conclusion of the project any minor contamination can be disposed of easily and effectively. For a larger scale operation, the plastic sheet is replaced by a shallow flat tray with raised edges which will serve to confine any accidental "spill". This type of precaution is also extended to many individual operations. For example, when evaporating a solution of medium or high specific activity in a small beaker under an infra-red lamp it is customary to stand it inside a petri dish.

Another very common type of precaution adopted when working above the millicurie level, is the use of thin rubber surgeons' gloves while carrying out manipulations. Many newcomers to the field of radiochemistry erroneously believe the use of the gloves is solely to protect the operator. This is not so. The gloves have a two-fold function: to protect the person using them, and also to protect other people working in the same laboratory from cross contamination. An error made by the improperly trained is to handle taps or inactive reagent bottles white still wearing the gloves, thereby contaminating these items which are subsequently used by others about any anticipation that they are contaminated. Gloves, when worn, should *only* be used whilst actually handling the active materials and should be removed before touching any other item of laboratory equipment , even if it means removing them for only 30 seconds. The correct method of removal is to peel them off from the wrist so that they come off inside-out with the active side inward, to be laid down on an inactive part of the bench. They are turned right side out immediately prior to re-use.

Two important points to be remembered are that precautions along the lines mentioned above are not substitutes for careful manipulation, but are a second line of defence in case of mishap, and that the key to safe radiochemical work is to anticipate all possible mishaps.

(C) Some Special Techniques

(a) Heating

Electrical methods of heating; hot plates, mantles and infrared lamps—particularly the latter—are preferred to gas. Infrared heat lamps are particularly useful for source preparation and evaporation of active solutions. Although evaporation is slower using lamps, the heat is supplied to the surface so "bumping" is eliminated.

(b) Measurement of Volume

Where possible, volumes are measured directly into calibrated centrifuge tubes in which reaction is to take place, or by the use of plunger pipettes. *Under no circumstances, are mouth operated pipettes used in a radiochemical laboratory*—for either active or inactive solution measurement. Where a titration is to be performed it is customary to measure the active solution with a plunger pipette and the inactive solution with a burette.

(c) Source Preparation

The measurement of γ, or high energy β activities, can be made with accurately measured volumes of solution in special liquid Geiger tubes or with scintillation counters. There are some limitations to this method particularly where a significant contribution to the count rate is made by β-particles. For example, the same container must be used when comparative determinations are made since small differences in geometry or wall thickness can materially affect β absorption efficiencies. The method is also unsuitable where high accuracy is required in the comparison of count rates between two different solvent media—*i.e.* in partition coefficient measurements—because the different density of the two media results in a different degree of β absorption. For all α and most β assay work, solid sources prepared by evaporation of an accurately measured aliquot on a planchet are employed. In preparing these sources, due on a planchet are employed. In preparing these sources, due recognition should be given to source self-absorption effects (Expt. 13) and back scattering effects (Expt.8).

(i) Sources from aqueous solutions. Micropipettes are available for measuring volumes down to 0.001 ml (1λ). Sources are generally prepared by evaporation of measured volumes of active solution on metal planchets. These are usually made of aluminium, though other metals—particularly stainless steel—are often preferable because of their greater resistance to chemical attack. Glass trays are also occasionally used.

Besides the different range of volumes which are measured by micropipettes, they also differ from conventional pipettes in that they are designed to *contain* and not to *deliver* the stated volume—a factor which influences the method of operation. A plunger, operated with a spiral motion between thumb and forefinger, is used to draw the solution about an eight of an inch above the calibration mark. Extreme care, especially with smaller pipettes is necessary to avoid overfilling. A paper tissue is used to wipe the outside and then held at the tip of the pipette to adjust the meniscus to the calibration mark.

The contents are then slowly ejected on to the centre of a planchet under an infra-red lamp. The first pipette rinse in the form of gobule of dilute acid or water drawn up from a small piece of plastic is also ejected onto the planchet.

Decontamination of the pipette should be effected before subsequent use with a least two dilute acid washes—the acid being drawn up well above the graduation mark, and finally two acetone washes. Air is then drawn through the pipette to dry it.

(ii) Organic solutions. The technique with organic solutions is similar to that employed with aqueous solutions except that some of the contents of the pipette will creep up the outside of the tip while the solution is being ejected so it is necessary to run a drop of solvent down the outside of the pipette after the contents have been ejected. Furthermore, it is usually necessary to place the planchet on a ring on a hot plate as well as using an infra-red lamp due to the lower volatility of the material.

(d) Active Waste Disposal

There are two philosophies concerning the disposal of radioactive waste. One is to *dilute* and *disperse* the material, the other is to *concentrate* and *contain* the material. Former, which is often applied to tracer work at the microcurie level, consists of flushing active solutions down the drain with a large volume of water or incinerating active tissues, etc. along with other inactive material and relying upon atmospheric dilution. The latter method involves the storage of active material in sealed containers for subsequent collection and disposal by some government agency. Because of storage difficulties, medium and high levels of activity are normally processed in minimum possible volumes.

Special Chemical Separation Methods

Chemical concentrations of radioisotopes are frequently extremely low—so low that the solubility product of many "insoluble" compounds, is very often not exceeded. Consequently these "insoluble" compounds, which might otherwise have been used to effect a separation, remain soluble.

Some idea of the quantities of material involved can be gauged from Table 11.3 which lists the mass of material in micrograms associated with one millicurie ($2.2 \cdot 10^9$ disintegrations per minute) of some of the more common radioisotopes. It will be seen for example, that one

Table 11.3
Activity of Some Common Radioisotopes

Radioisotope	*Half-life*	*Activity (micrograms per millicurie)*
^{24}Na	15.06 hours	$1.15 \cdot 10^{-4}$
^{32}P	14.3 days	$3.5 \cdot 10^{-3}$
^{35}S	87.1 days	$2.34 \cdot 10^{-2}$
^{36}Cl	$4.4 \cdot 10^5$ years	$44.1 \cdot 10^3$
^{42}K	12.4 hours	$1.66 \cdot 10^{-4}$
^{45}Ca	152 days	$5.23 \cdot 10^{-2}$
^{51}Cr	27.8 days	$1.08 \cdot 10^{-2}$
^{59}Fe	45.1 days	$2.03 \cdot 10^{-2}$
^{60}Co	5.27 years	$8.81 \cdot 10^{-1}$
^{64}Cu	12.8 hours	$2.61 \cdot 10^{-4}$
^{82}Br	35.8 hours	$9.37 \cdot 10^{-4}$
^{90}Sr	19.9 years	4.99
^{131}I	8.14 days	$8.15 \cdot 10^{-3}$
^{137}Cs	33 years	$1.26 \cdot 10^1$
^{144}Ce	282 days	$3.1 \cdot 10^{-1}$
^{226}Ra	1622 years	$1.0 \cdot 10^3$

microcurie of strontium-90 would involve a mass of only $5{\cdot}10^{-9}$ g. This in say, 10 ml would constitute a chemical concentration of just over $5{\cdot}10^{-9}$ molar, or in other words 1/20th the concentration of H_3O^+ or OH^- ions in pure water.

Of the many special chemical methods that have been developed or adopted in radiochemical work only the three most important will be treated here. These are carrier precipitation, ion exchange, and solvent extraction.

(A) Carrier Precipitation Method

The addition of either an inactive isotope of the same element as the radioisotope, or of an element having similar chemical properties, to a solution prior to performing a precipitation—in order to "carry" a particular radioisotope in a precipitate—is well established practice. For example, the use of barium to carry down radium in a sulphate precipitate. Carriers are also sometimes added to a solution to prevent the adsorption of a radioisotope on a precipitate when it is required that the radioisotope in question should remain in solution. For example, in the absence of a carrier, group one elements will frequently be adsorbed on gelatinous hydroxide precipitates. Carriers used for this purpose are termed "hold-back carriers".

Carrier techniques have the advantage that they are a modification of a common chemical method and the simple laws of solubility apply. The main disadvantage is that it is sometimes necessary to recover the radioisotope in high specific activity by separating it form the carrier. With isotopic materials this is not possible, while with an element of very similar chemical properties, it is often very difficult.

(B) Ion Exchange and Solvent Extraction Methods

Ion exchange and solvent extraction methods for the separation of elements over a concentration range from the tracer level to molar concentrations have been developed to a high degree of efficiency over the past thirty years. Ion exchange procedures—in common with solvent extraction procedures—rely for their operation upon the type of ions present in solution.

(a) Complex Formation in Aqueous Media

The formation of soluble complexes of such elements as silver and copper when excess ammonium hydroxide is added to solutions of the slats of these elements is well known to students. The wide range of complexes of such elements as cobalt is also covered in many text books on inorganic chemistry, and the analytical chemist is well acquainted with many organic complex forming reagents. Not so well appreciated are the simplest and most common of all complexes—the ionic complexes formed by a very large number of metal ions with mineral acid anions. For example, a solution of ferric chloride in water does not consist of simply Fe^{3+} and Cl^- ions. On the contrary, in a 0.1 molar solution—without excess acid present—less than half of the iron exists as Fe^{3+}. The remainder is present mainly as a series of complexes of the type $FeCl^{2+}$, $FeCl_2^+$ and $FeCl_3^0$. The proportion of higher complexes which are formed by reactions of the type:

$$Fe^{3+} + Cl^- \rightleftarrows FeCl^{2+}.$$

$$FeCl^{2+} + Cl^- \rightleftarrows FeCl_2^+ \text{ etc.}$$

is increased by hydrochloric acid. Complexes are not restricted to cation and neutral types, in fact iron and very many other metals form anionic complexes; for example, the sulphate complex of uranium formed during uranium extraction from ores by sulphate leaching, and the complex $Th(NO_3)_6^{2-}$ employed in the thorium-scandium separation described in Expt. 16.

The utilisation of ion exchange and solvent extraction separation techniques depends upon an appreciation of the complex forming ability of metals. An account of the individual complex formation ability of all metals in various acid media is far beyond the scope of this book. All that can be given is a broad generalization to act as a guide for potential applications of ion exchange and solvent extraction. As a general rule, metals in groups 1A and 2A of the Periodic Table exist as cations in solutions of mineral acids, while metals in the groups 1B, 2B and those exhibiting a valency of 3 form neutral complexes reluctantly and in a very few cases anionic complexes. Metals exhibiting valencies 4,5 and 6 form neutral and anionic complexes readily. The higher the valency, the larger is the variety of anions with which a metal will complex. Detailed information concerning the actinides is given in Chapter 9.

(b) Ion Exchange

Ion exchange can be classified into two classes, *cation exchangers* and *anion exchangers*. Cation exchangers adsorb and desorb (*i.e.* exchange) cations and cationic complexes, whilst anion exchangers adsorb and desorb anions and anionic complexes. Although many inorganic compounds show ion exchange properties, the most widely used materials are synthetic resins. These are polystyrene based polymers with functional sulphonic acid groups incorporated in the structure (cation resins), or with incorporated quaternary amine groups (anion resins). Some commonly used resins are ZeoKarb 225 and Dowex 50 (cation) and DeAcidite FF and Dowex 1 (anion).

Ion exchange columns comprise a tube with a large length to diameter ratio into which the resin is slurried. The columns are always kept wet. They are frequently surrounded by a vapour jacket for more efficient operation at elevated temperatures (Fig. 11.2).

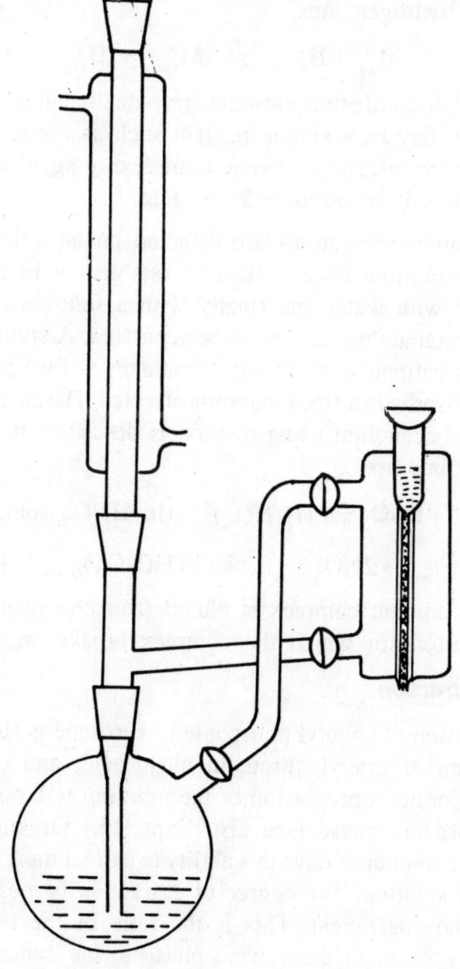

Fig. 11.2. Jacketed ion exchange column for operation at elevated temperatures.

Cation resins are most frequently—though not exclusively —used in chromatographic types of separation, while anion resins are more commonly used in clear cut anion—cation separations. In cation chromatographic separations, the resins is first treated with strong acid to ensure that it is in the hydrogen ion form, then with water and finally with a solution of composition similar to that containing the cations to be separated.

A small volume of solution containing the cations is then placed on top of the resin column where they are adsorbed, with simultaneous desorption of hydrogen ions.

$$M^{+}_{aq.} + H^{+}_{resin} \rightleftarrows M^{+}_{resin} + H^{+}_{aq.}$$

The adsorbed cations are then chromatographically eluted with a suitable solution which may be a simple reagent such as concentrated HCI or may take the form of a more exotic complexing agent as for example the ammonium salt of an α-hydroxy acid.

Anion-cation separations are effected on an anion resin which has been preconditioned with alkali to establish it in the OH^{-} form, then contacted with water and finally with a solution of composition similar to that containing the ions to be separated. A solution containing the mixture of cationic and anionic complexes is then passed through. The latter are absorbed and the separation effected. The clear cut separation of thorium and scandium using method is described in Expt. 16. The relevant equations are:

$$TH^{4+} + 6NO_3^- \rightleftarrows TH(NO_3)_6^{2-} \text{ (in } HNO_3 \text{ soln.) and}$$

$$TH(NO_3)^{2-}_{6(aq.)} + 2NO^{-}_{3(resin)} \rightleftharpoons TH(NO_3)^{2-}_{6(resin)} + 2NO^{-}_{3(aq.)}$$

The adsorbed anionic complex is eluted from the resin by providing solution conditions in which the complex breaks up.

(c) Solvent Extraction

If a solution of tributyl phosphate in kerosene is shaken or stirred with a solution of uranyl nitrate in nitric acid, and the two phases allowed to separate, a proportion of the uranium will be found to have entered the organic phase (see also Expt. 15.) Organic solutions of many organic compounds have this ability to extract inorganic complexes from aqueous solution. The degree of extraction is measured in terms of the *partition coefficient*. This is the ratio of the concentration of the extracted species in the solvent phase to the concentration in the aqueous phase,

$$\text{Partition coefficient (K)} = \frac{\text{concn. org. phase}}{\text{concn. aq. phase}}$$

Partition coefficients for the extraction of a very large number of metals by a variety of solvents under various solution conditions are to be found in the chemical literature. The values vary from zero to tens of thousands. In choosing a solvent for a particular separation process, two things must be borne in mind.

(a) A high partition coefficient will provide the greatest recovery with the minimum number of extractions.

(b) The greatest difference between the partition coefficients for two metals to be separated will lead to the most efficient separation.

In practice, it is best to aim for the greatest difference in coefficients and make up for low numerical values by using multiple stage extraction.

The factor which influences the type of solvent suitable for a particular extraction is the ionic species present in solution. For example,

(*i*) *Cations and cationic complexes* are extracted by acidic solvents such as dibutyl phosphoric acid (DBP) and di-2-ethylhexyl phosphoric acid.

(*ii*) *Neutral complexes* are extracted by many ethers, esters, ketones and some high molecular weight alcohols. Particularly suitable for this class of complex is tributyl phosphate (TBP).

(*iii*) *Anionic complexes* are extracted by many basis solvents—in particular tertiary amines such as trioctylamine (TOA).

Sources of Experimental Error

Assuming perfect instrumentation and manipulation there are three sources of counting error which can occur and for which allowance should be made. These are:

(i) Background. Radiation from cosmic and terrestrial origin as well as the immediate surroundings will be present. In experiments it is necessary to take precautions to ensure that the background is as low and constant as possible. Where necessary the background count should be determined and subtracted from the observed count should be determined and subtracted from the observed count to arrive at the true count.

(ii) Statistical Errors. Radiations are emitted from a source at random intervals, not at uniformly spaced intervals. Because of this, successive counts on the same source under identical conditions will not be the same. The probable error associated with a count may be taken as the square root of the gross count recorded, for example, as 10,000 ± 100 (see Expt. 2 on "Statistics of Radioactive Measurement").

(iii) Coincidence Errors. Due also to the random nature of decay and the non-uniformity of the intervals between pulses, there is a finite probability that more than one pulse will arrive during the response time of the counter. For low count rates, this is negligible but for high count rates, large errors can be introduced by neglecting this factor (see Expt. 5 on "Determination of Dead Time").

As a general rule, in cases of low count rate coincidence losses will be negligible but background counts will be of utmost importance; while in cases of high source strengths the background count will be negligible nut coincidence losses will be significant. Also, to obtain low statistical errors at low count rates, it is necessary to count for extended periods of time before dividing the gross count by the time interval to obtain the count rate.

SELECTED EXPERIMENTS

The following experiments are included to further emphasize particular topics included in the earlier sections of the book, or to demonstrate special techniques.

The first experiment is qualitative and is intended as a demonstration of the properties of nuclear radiations. The remaining experiments are of a quantitative nature.

Individual experiments are not intended to involve the same fixed period of time although each one can be carried out within a single three hour session. In many cases they may be used in combination with one another.

Experiment 1

Properties of Nuclear Radiations

For this experiment, the following equipment an sources are recommended.

(i) A thin end-window Geiger detector connected to a radiation monitor by a flexible lead.

(ii) A 0.05 to 0.1 μc α source; for example, ^{233}U, ^{239}Pu, ^{230}Th or ^{241}Am

evaporated on a stainless steel planchet and "flame sealed" by raising the planchet to red heat in a bunsen flame.

(iii) A 0.05 to 0.1 μc β source; for example, ^{90}Sr or natural uranium*. Uranium is weak α emitter, but the second daughter, ^{234}Pa, which is present in equilibrium with the uranium is a high energy β emitter.

(iv) A 5 μc γ source of ^{60}Co*.

(v) A 5 μc collimated β, γ source prepared by evaporation of a ^{137}Cs solution at the bottom of a $^3/_{16}$" diameter, 1" deep hole in an aluminium block.

(i) Random Nature of Decay. With any one of the sources positioned so as to give a count rate of the order of 100 counts per minute the non-uniformity in the spacing of the pulses recorded will be apparent. It can also be shown that successive counts taken under the same conditions are not identical.

(ii) Range of α-, β- and γ-radiation. The very short ranges of α-particles compared with β- and γ-radiation can be shown by placing the detector immediately above the α source and observing the count rate as the detector is slowly moved away. Besides a steady decrease in count rate due to the inverse square law effect, there will be an abrupt drop at about 2 cm. In the case of β and γ sources, there will only be the steady fall in count rate caused by the inverse square law effects.

(iii) Penetration of α-, β- and γ-radiation. By placing, in turn, the α, β and γ sources 1 cm below the detector window, it can be readily shown that:

(1) α-particles can be stopped by aluminium sheet, aluminium foil and even by air-mail weight paper.

(2) β-particles are unaffected by paper and foil, but are attenuated by thin aluminium sheet (20 to 24) gauge) and are completely stopped by thick aluminium sheet (10 to 12 gauge). The relative effect of different thicknesses of aluminium sheet is a direct illustration of the principle of the β transinission thickness gauge.

(3) γ-radiation is unaffected by all of the above-mentioned materials and not even thin lead sheet has any appreciable effect.

* Safe and convenient β and γ sources can be prepared by incorporating a suitable compound of the isotope concerned—usually the oxide—as the filler in an epoxy resin mixture which is then cast into planchets and allowed to set. These sources are also suitable for use in Expts. 2,3,4,5,9 and 10.

(iv) Deflection in a Magnetic Field. The collimated β, γ source should be placed 5 to 7 cm below the detector and the count rate observed. If a small pocket horseshoe magnet is placed immediately above the hold in the block the β-particles will be deflected and only the γ count will be recorded. By moving the detector in an arc over the source the deflected β-particles may be located, and if the direction of the field is known the nature of the charge on the β-particles may be inferred.

(v) Scattering of β-particles. The collimated β source and detector should be arranged close together with their axes at right angles to one another. A piece of lead—or other material of high atomic number—may then be used as a "mirror" to deflect a large number of β-particles through 90° into the detector.

Experiment 2

Statistics of Radioactive Measurement

The random nature of radioactive decay may be observed simply by watching counts "come up" on a scaler, when it will be apparent that the time interval between successive counts is not constant. A source with an average count rate of 1000 counts per minute may give 980 counts in one minute, 1100 in the next and so on. If, for a very large number of determinations, the frequency with which an observed count is obtained is plotted against that count a Normal (or Gaussian) statistical distribution is obtained. In such a distribution 68% of all measurements should lie between ± one standard deviation from the mean and 95% between ± two standard deviations from the mean.

The aims of this experiment are to illustrate the random nature of radioactive decay, to show that a large number of successive counts approximate to a Normal statistical distribution and to establish that the standard deviation determined from a large number of counts is, in fact, approximately equal to the square root of the mean of the observed counts. Here-after, in subsequent experiments, it will only be necessary to take a single count and to quote the observed count "n" as $n \pm \sqrt{n}$ in order to indicate the reliability of the recorded count.

In this experiment, a scaler connected to any type of detector is used together with a radioactive source of sufficient strength to give approximately 10^4 counts per minute. A minimum of 30 separate successive one minute counts are to be taken under identical conditions. These counts should be tabulated with their deviations from the mean and the square of the deviations. The root mean square deviation (standard deviation) is then determined and compared with the square root of the mean. Draw a histogram, plotting the number of counts which appear in a certain interval against count intervals. Compare the shape of the outline of the histogram with a Normal distribution and record the number of counts which lie within one standard deviation from the mean and the number that lie within two standard deviations from the mean.

Experiment 3

Geiger Tube Characteristics

The Geiger tube consists of a cylindrical cathode with a central wire anode and is fiiled with an inert gas and quenching agent to a total pressure of about 10 cm of mercury. A potential is applied between the anode and cathode sufficient to cause multiple ionisation by collision processes after primary ionisation is produced by the radiation. If the potential applied to a Geiger tube is gradually increased from zero, a threshold potential will be found above which the tube begins to operate. Usually within 25 volts of the potential the "Geiger plateau" is reached where the number of counts recorded is almost independent of the applied potential.

Occasionally a Geiger tube must be replaced. The method used to determine its condition and also its correct operating potential is to determine its plateau characteristics. In general, the plateau should be over 200 volts long, with a slope of less than 0.05% per volt. The correct operating potential is normally at a point 50 V above the threshold potential on the plateau where voltage supply variations will have little effect on the count rate.

The aim of this experiment is to determine the plateau characteristics and correct operating potential of a Geiger tube. To do this, place a radioactive source of sufficient strength to give approximately 10^4 counts per minute below an end-window Geiger tube, or place a β active solution of similar strength in a liquid Geiger tube assembly. Switch the instrument on to the "count" position and slowly increase the EHT form its minimum position until a count is observed. Reduce the potential 25 volts and take 2 minute "counts" from this position, increasing the EHT by 20 or 25 volts after each determination.

Graph the result on linear graph paper with the count rate as the ordinate and the EHT potential as the abscissa. Continue to plot values for 250 volts beyond the threshold potential or until the count rate begins to increase above its constant plateau value—whichever happens first. An excursion into the continuous discharge region can seriously damage tube.

Experiment 4

Scintillation Counter Characteristics

A scintillation counter consists of:

(i) A phosphor containing molecules which, when excited by radiation, return to their ground state with the emission of light.

(ii) A photomultiplier. Light emitted from the phosphor ejects photoelectons from the cathode of the photomultiplier. These electrons, accelerated by a potential of some 100 volts to the first of many dynodes, produce secondary electrons which in turn, are accelerated to the second dynode and so on. The total electron amplification of the photomultiplier is a function of the applied potential.

(iii) A recording section. Pulses from the photomultiplier are first fed through a small low gain amplifier, usually housed in the photomultiplier assembly, then through a discriminator which can be set to pass only those pulses which exceed a certain amplitude to the recording section. The magnitude of the pulses arriving at the discriminator is a function of the size of the light flashes in the phosphor and of the subsequent amplification in the phosphor and of the subsequent amplification in the photomultiplier, which in turn is a function of the potential applied to it.

When the count rate is plotted against the potential which is applied to the photomultiplier, it is found to increase with applied potential, but the curve obtained has an inflection where the count is not greatly influenced by applied potential.

This curve is in effect the composite of two separate curves, viz,:

(i) An increase in potential causes an increase in the amplification of the pulses, so that more and more of them exceed the discriminator bias setting, and are therefore recorded. In theory this would result in a curve where, initially, the count rate increased as the potential increased and should reach a point where all the pulses are recorded. The count rate should then be independent of further increases in potential.

(ii) An increase in potential also causes an increase in the "noise" pulses in the photomultiplier tube. The count rate attributable to this phenomenon increases rapidly with an increase in the applied potential, particularly at high values.

The aim of this experiment is to study the characteristics of a scintillation counter by studying the variation in count rate from a source as a function of discriminator and high tension settings, and thereby to determine suitable operating conditions.

With the discriminator set at 5 volts and using a radioactive source capable of giving about 10^4 counts per minute, determine the count rate as a function of EHT potential from 700 volts at intervals of 50 volts. Remove the source and carry out the same procedure for any one of the previously used discriminator settings to determine the background count characteristics.

Using semi-log graph paper, plot count rate on a logarithmic scale against the potential applied to the photomultiplier for each value of the discriminator setting. Select suitable operating conditions where the count rate is not unduly dependent upon high tension or discriminator potentials. Explain the reason for the horizontal displacement of the curves as the discriminator bias is increased.

Experiment 5

Counter dead-time

Counters have a recovery time—or "deed-time" after each recorded count during which they are insensitive and will not record a subsequent count. At high count rates the observed count rate is considerably lower than the true count rate and a correction must be applied. The dead-time be due to either the detector and/or the electronic circuitry.

If *R* is the observed count rate per unit time, and *t* is the dead-time after each count, the total dead-time per unit time = *Rt*.

If *N* is the true number of counts per unit time, the number of counts lost due to the dead-time = *N* — *R*. But this is also equal to the true count (*N*) × fraction of time that the counter is insensitive (*Rt*).

$$N - R = NRt$$
$$R = N(1 - Rt)$$

or

$$N = \frac{R}{1 - Rt} \quad (1)$$

From this it is seen that *t* may be determined if *R* and *N* are known. As a general rule, however, *N* is not known.

If R_1 is the observed count rate for a source with a true count rate N_1 and R_2 is the observed count rate for a source with a true count rate N_2

Then :

$$N_1 = \frac{R_1}{1 - tR_1}$$

and
$$N_2 = \frac{R_2}{1 - tR_2}$$

Therefore
$$\frac{N_1}{N_2} = \frac{R_1 - tR_1R_2}{R_2 - tR_1R_2}$$

Thus, only the ratio N_1/N_2 need be known as well as the values of R_1 and R_2 in order to determine t.

Method. Two almost identical cobalt-60 sources, each approximately 5 microcuries, are used on either side of a Geiger tube to perform this experiment.

One of the two sources is first placed in the *fixed* position on one side of the tube and its count rate determined over an extended period of time (10 min.). This source is then removed and the second source placed on the other side of the tube in the adjustable holder. The position of this source is then varied so as to give the same count rate as was given by the first. The first source is then replaced in its original position to effectively double the source strength.

If we now assume that the source strength of each of the two individual sources is N_1 and that the combined strength is N_2, then;

$$\frac{N_1}{N_2} = \frac{1}{2} \tag{2}$$

Note that the observed count for both sources together (R_2) will not be quite twice the individual counts of each separate source, which should be averaged to obtain R_2.

Calculate and record the dead-time.

In subsequent experiments the true count can be determined from the observed count by using eqn. (1) above if the dead time is known. In the case of equipment with dead-times of either 300,400 or 500 microseconds, the corrections to be added to the observed count can be readily determined from Fig. 11.3.

Experiment 6

The inverse square law and collaboration against a radium standard

Measurements obtained with each of the following instruments:

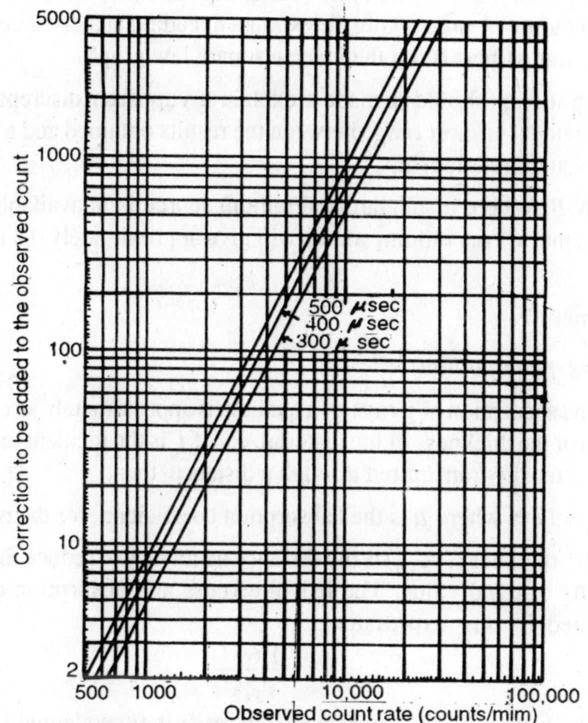

Fig. 11.3. Dead-time corrections as a function of observed count rate for instruments with a dead-time of 300, 400 and 500 microseconds.

(1) a general purpose laboratory monitor fitted with a tubular Geiger tube, and

(2) a portable ion chamber type of radiation monitor should be recorded at 10 cm intervals (up to 1 meter) between each of these instruments and a standardized radium source of about 1 mc. Distances measured in centimeters should be taken between the radium source and the axis of the Geiger tube in the case of the Geiger operated monitor and between the source and mid point of the ion chamber in the case of the ion chamber radiation monitor.

The results should be tabulated and the count rates obtained should be plotted against 1/(distance)^2 on linear graph paper to verify the inverse square law. The count rate on the Geiger operated monitor corresponding to 1 mr/b should be recorded.

The calibration of the other instrument should be checked against the standard radium source taking the estimated reading at a distance of 1 meter from the line of best fit on the inverse square law graph.

An attempt should be made to explain any apparent discrepancy which may be evident at short range between the results obtained and a theoretical inverse square relationship.

(*N.B.* Where a standardized radium source is unavailable it can be assumed that a 1 mc radium source will give approximately 0.8 mr/h at one meter).

Experiment 7

Shielding-γ attenuation

The absorption of γ-rays by a medium is approximately an exponential function of the thickness of the material, *e.g.* if I_0 is the incident intensity and I_d is the intensity transmitted through a distance then:

$I_d = I_0 e^{-\mu d}$ where μ is the "absorption coefficient" for the material.

The "half-thickness" ($d_{1/2}$) is the distance necessary to reduce the intensity to half its original value. The half-thickness and absorption coefficient are related by the expression:

$$\mu = \frac{0.693}{d_{1/2}}$$

The purpose of this experiment is to verify the exponential relation and to determine the absorption coefficient and half-thickness for brass and iron using absorbers of various thicknesses.

A suitable source consists of a shielded 1 curie ^{137}Cs or, ^{60}Co source in front of which is a lead brick with a small circular aperture to provide a collimated beam of γ-rays. A second aperture is located immediately in front of a Geiger detector. The absorbers to be studied are placed on a cradle between these two apertures.

Ensure that the source, apertures and detector are aligned and then determine the transmitted γ intensity (I_d) for the various thicknesses of iron and brass from 1 cm to 7 cm by using various combinations of 1,2 and 4 cm absorbers. Note that the duration of the count will have to be extended as the thickness increases and the count rate decreases in order to obtain a reasonable order of accuracy.

Plot the results on semi-logarithmic graph paper with the count rate on the log scale. Verify the exponential relation and calculate the half thicknesses and absorption coefficients for the materials supplied.

What relation, if any, exists between the physical properties of the absorbing materials and their absorption coefficients?

Experiment 8

β *scattering*

β–particles are readily scattered and a significant fraction of β-particles striking a material will be scattered back (or, as we say, back-scattered). The object of this experiment will be to study the effect of various thicknesses of aluminium on back-scattering, and the effect of materials of different atomic number on saturation back-scattering.

(1) A special source consisting of an annular deposit of ^{90}Sr around a central $^3/_4$" diameter hole in a piece of thick aluminium, should be placed approximately 2 cm below an end-window Geiger tube with the source side facing downward. Various β reflectors are placed, in turn, a further 2 cm below the source to scatter the β-particles back through the hole into the detector.

Using aluminium of various thickness from 5 mg/cm^2 to 500 mg/cm^2, plot back-scattered radiation against thickness. Using thick absorbers of different materials (aluminium, copper, iron, cadmium and lead) plot the saturation back-scattering against the atomic number.

(2) using 100λ of a ^{137}Cs solution (approx. 0.5 μc/ml), prepare two identical β sources, one on an aluminium planchet and the other on a stainless steel planchet. Count each of the sources. Note and explain any significant difference in the count rate.

Experiment 9

β*-particle energy by aluminium absorption measurements*

β-particles from a radioactive isotope have a distribution of energies which is characteristic of the particular isotope concerned. It is used to determine the maximum energy of the β-radiation in order to identify the isotope.

One of the methods commonly used to determine the value of E_{max} is to study the attenuation of β-particles in aluminium absorbers. E_{max} can then be determined by one of several methods; that described here is that of Bleular-Zunti.

Between the β source to be studied and a thin end-window Geiger detector, place aluminium absorbers ranging in thickness from 5 mg/cm^2 to 400 mg/cm^2. The count rate is observed for varying thicknesses of aluminium, the duration of the count being progressively increased to maintain a reasonable counting accuracy.

By plotting the relative count rate as a function of absorber thickness on semi-log paper, the thickness d_n required to reduce the count by 2^n (n= 1,2,3,...) can be found and E_{max} determined by reference to the curves in Fig. 11.4 The experimental absorption curve obtained should approximate to a straight line on a semi -log plot, flattening out at the end to a constant level corresponding to the γ background. Should a curve result, consideration should be given to resolving the curve into two straight lines each corresponding to one of two different β-particle energies.

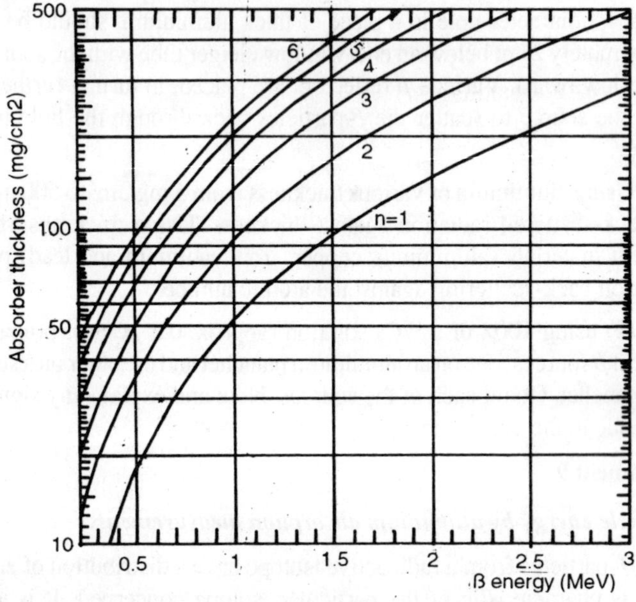

Fig. 11.4 : β energy determination by the Bleuler–Zunti method. Curves related β energy to the absorber thickness required to reduce the count by a factor 2n, where n = 1, 2, 3. etc.

The various values of E_{max} obtained for n=1,2,3, etc. should—with the occasional exception of n = 1— be approximately constant, and correspond to the maximum value of the energy of the β-particles.

Corrections for dead-time of the counter, and absorption by the air gap and Geiger window should be made. The density of air may be taken as 0.0012 g/cc.

Experiment 10

γ spectrometry—single crystal, single channel spectrometer

The γ-radiation emitted by a nuclide has a characteristic discrete energy (or set of energies) and the γ spectrum is a convenient way of identifying a nuclide.

For γ spectrometry, a phosphor with a high probability of absorbing the incident radiations by photoelectric absorption must be chosen, as the results in all the photon energy being converted into light flashes with an intensity that is proportional to the energy of the absorbed photons. Measurement of the magnitude of the light flashes thus provides a method for the determination of γ spectra.

A single channel spectrometer consists of:

(1) A scintillation head comprising a thallium activated sodium iodide crystal and photomultiplier.

(2) A linear amplifier.

(3) A pulse height analyser which is used to determine the number of pulses which lie between two preset values, and which is used to scan the pulse spectrum.

(4) A ratemeter.

The purpose of the experiment is to illustrate a method for the determination of a γ spectrometer calibration curve for a given set of EHT potential and amplifier instrument settings.

To ensure that the instrument is operating properly, place a ^{60}Co source of approximately 5 to 15 μc below the crystal and, with the analyser, scan the spectrum backwards, beginning at 50 volts. About half-way down the scale, two adjacent ^{60}Co peaks should be resolved if the EHT and amplifier settings are correct. Record the instrument settings and then proceed to scan the spectrum from 5 to 50 volts with a channel width of 1 volt.

Replace the ^{60}Co with a ^{22}Na source of similar strength and again scan the spectrum. In each case plot the count rates against the analyser potentials on linear graph paper and by reference to a Chart of the Nuclides identify the γ energy of the spectrum peaks, remembering also that the ^{22}Na will have a strong γ peak component at 0.51 MeV corresponding to positron annihilation. From these results construct a γ energy *vs.* analyser potential calibration curve for the instrument.

Using a ^{137}Cs source, and with reference to the calibration curve, determine the γ energy of ^{137}Cs.

Experiment 11

Neutron activation and half-life of a radioactive isotope

In each of these experiments the use of either a radium-beryllium or an americium-beryllium neutron source is necessary.

(i) Iodine

Radioactive isotopes may frequently be obtained by neutron irradiation of natural stable isotopes, *e.g.*

$$^{127}_{53}I(\text{stable}) + {}^{1}_{0}n \rightarrow {}^{128}_{53}I \xrightarrow{\beta} {}^{128}_{54}Xe$$

An irradiated liquid sample of ethyl iodide should be removed from a neutron source after irradiating for 2 hours, and transferred to either a liquid Geiger counter or a liquid (well scintillation counter. The count rate should be determined over time intervals of 1 minute, with 1 minute between each determination, up to a total time of approximately 50 minutes. Correct the observed counts for dead-time and plot the count rate on a logarithmic scale against time plotted on a linear scale. The half-life may then be determined from the slope of the line obtained.

(ii) Rhodium

An irradiated solid rhodium source should be removed from the neutron source after a 30 minute irradiation, transferred as rapidly as possible to a solid source Geiger assembly and the count determined over time intervals of 10 seconds with 10 seconds between each determination up to a total time to 6 minutes. Plot count ate corrected for dead-time, against time on semi-log graph paper. On this occasion a curve will be obtained which is the composite of two straight lines, each representing the decay characteristics of each of two isomers.

Experiment 12

Absolute Counting-efficiency of β detectors

The object of this experiment is to determine the absolute activity of β active ^{137}Cs solution and to estimate the efficiency of a liquid Geiger tube.

To determine the absolute activity of solution a source is prepared by evaporating 0.1 ml of stock ^{137}Cs solution* and its activity is measured with

* The solution should be similar to that used into the β scattering experiment, viz. 0.5 mc/ml.

an end-window Geiger tube in a perspex stand under conditions of known geometry. It is assumed that all β-particles entering the Geiger tube are counted and it only remains to determine what proportion of the β-particles emitted by the source enters the tube. Factors to be considered are:

(1) Geometrical efficiency of the arrangement *i.e.* the proportion of the β-particles which travel within the solid angle subtended by the aperture in front of the tube window. The ratio of β-particles entering the detector to all those emitted, will be the same as the area of the surface of a sphere of radius equal to the distance from the source compared with the area of the aperture.

(2) Absorption of some of these β-particles by the air and Geiger tube window. This may be determined by placing absorbers between the source and tube to obtain an absorption curve and extrapolating this curve back to zero thickness. Density of air = 0.0012 g/cc.

(3) Back scattering from the material on which the source is prepared. The source should be prepared on thin aluminium foil to render this effect negligible.

(4) The background and γ component of the source which may be determined by placing a thick aluminium absorber between the source and the detector.

To determine the efficiency of a liquid Geiger tube, the calibrated source is dissolved in dilute hydrochloric acid, and the solution made up to 100 ml. Ten ml of this solution should be counted in a liquid Geiger tube.

Experiment 13

Self-absorption

When increasing thicknesses of aluminium are placed between a β source and a detector, the count rate decreases due to absorption, Similarly when β sources of a finite thickness are counted, the count rate is reduced by absorption in the source material itself. This is known as self absorption in the source material itself. This is known as self absorption and must be allowed for in all accurate radiochemical determinations. It is extremely important in the case of low energy β emitting and α emitting nuclides, of lesser importance in high energy β emitters and of little consequence when a γ emitter is assayed by γ detection methods.

The purpose of this experiment is to illustrate self absorption by preparing and counting several sources each containing the same amount of ^{35}S—a low energy β emitter (0.167 MeV), in varying amounts of source material (barium sulphate) and to compare, graphically, the count rate against source thickness in mg/cm^2.

Method. Varying known quantities of barium sulphate, all containing the same amount of ^{35}S, are prepared by precipitating barium sulphate from a solution containing a known quantity of sulphate ion using an excess of barium chloride solution. These precipitates are filtered in a demountable perspex filter to provide sources of known area to be mounted on an aluminium planchet and counted.

In each of ten 50 ml beakers add approximately 10 ml of water and exactly 1 ml of sulphur-35 stock solution (approx. 0.05 μc/ml). Place 1 ml of 0.05 $M H_2SO_4$ in the first beaker, 2 ml of 0.05 $M H_2SO_4$ in the second and so on. Warm the solutions. Precipitate the sulphate by the careful addition of excess barium chloride using approximately 1 ml 0.1 $M BaCl_2$ in the first, 2 ml of 0.1 $M BaCl_2$ in the second and so on. Filter, and while still damp, transfer the precipitates to source trays. Dry them under an infra-red lamp until the papers just start to curl and count in the top shelf of a Geiger assembly.

Experiment 14

Szilard-Chalmers Reaction

This experiment illustrates the formation of a radionuclide by the (n, γ) reaction and the chemical effect of the nuclear transformation—the Szilard—Chalmers reaction.

Ethyl iodide (5-10ml) is placed in a polythene tube fitted with a bark cork, and activated in a neutron source for approximately 2 hours. A small proportion of the ^{127}I will capture neutrons to become ^{128}I:

$$C_2H_5\ {}^{127}I + n \rightarrow C_2H_5\ {}^{128}I^* \xrightarrow{\text{De-excitation}}$$

De-excitation of the product by γ emission causes the residual nucleus to recoil and this recoil energy is frequently sufficient to break the chemical bond between this atom and the rest of the molecule. Recoiling ^{128}I atoms take part in many reactions with their surroundings but, in the end, about 50% can be separated as iodine plus iodide ion. Important consequences of this reaction are that one may expect to find the product of a neutron activation in a different chemical form to that of the target material, and that the reaction affords a means of isolating an irradiation product in high specific activity from a large mass of inactive isotopic material. A requirement is that exchange between product and parent species does not occur.

To measure the proportions of "organic" and "inorganic" iodine-128 present, transfer the irradiated ethyl iodide to a 40ml centrifuge tube, add an equal volume of aqueous sodium thiosulphate solution and stir vigorously with an electric stirrer for two minutes. Separate the phases and determine the activity in each phase, using a plunger pipette to withdraw 5 ml of each phase for successive counting in the same liquid Geiger tube. Note the time each phase for successive counting in the same liquid Geiger tube. Note the time each count was begun. Correct both counts for background and dead-time and determine what the second count would have been at the time of the first count allowing for radioactive decay (see Expt. 11 on "Half-life").

Report your answer as percent "inorganic" ^{128}I.

Experiment 15

Solvent Extraction

Determination of the distribution coefficient of Th between nitric acid and tributyl phosphate

In aqueous solution a large number of metal ions form complexes even with such simple anions as sulphate, halide and nitrate. The complexes may be cationic, neutral or anionic and can be extracted from aqueous solution by a suitable organic solvent. The concentration of the metal in the solvent phase divided by the concentration in aqueous phase after extraction is known as the partition coefficient.

The purpose of this experiment is to illustrate the existence of complex ions of thorium in nitric acid solutions, to illustrate the principle of solvent extraction, to introduce the technique of using tracers to "spike" a solution and to provide experience in preparing sources from both aqueous and organic solutions. The aim will be to determine the distribution coefficient for the extraction of thorium from HNO_3 into 20% (v/v) tributyl phosphate (TBP) in kerosene.

Pre-equilibrate about 2 ml of 20% TBP with twice its volume of 2 *M* nitric acid by stirring in a 15 ml centrifuge tube for 2 minutes. Place exactly 1 ml of 2 *M* nitric acid in a second 15 ml centrifuge tube, spike with 0.01 ml of ^{230}Th stock solution using a micropipette and then add exactly 1 ml of the pre-equilibrated 20% TBP. Stir Vigorously for 2 minutes with an electric stirrer, separate the phases in a centrifuge and prepare two duplicate sources from each phase on stainless steel planchets, using 0.1 ml for each source. Count the sources and calculate the partition coefficient.

Repeat the experiment using 3 *M* nitric acid and report the effect of the nitric acid concentration on the partition co-efficient.

Experiment 16

Separation of thorium and scandium by ion exchange

This experiment is intended to demonstrate the application of ion exchange separation methods and tracer techniques to follow the effectiveness of such a separation. The resin used is De Acidite FF, an anion exchange resin which has the ability to exchange and absorb anions but not cations. The column should be of the jacketed type (Fig. 11.2) and have a bore of 2 mm and a length of approximately 5 to 6 cm.

a) Column preparation

(1) Ensure that the column is packed with De Acidite FF anion exchange resin that is wet and free from air pockets.

(2) Heat the column by refluxing carbon tetrachloride vapour through the jacket.

(3) Pass approx 10 drops of 5 *M* nitric acid through at a flow-rate of approx 1 drop per 2 min. Applying external pressure (up to a max. limit of 10 cm Hg) if necessary.

(b) Separation

(1) As the meniscus of the acid reaches the top of the resin bed, thorium-230 (α) and scandium-46 (β) tracers in one drop of 5 *M* nitric acid should be slowly absorbed on the top of the column without pressure.

(2) When the meniscus of the spiked drop reaches the resin bed add one drop of 5 *M* nitric acid.

(3) When the meniscus of the drop of acid reaches the top of the resin a further 5 drops of acid are run through the column.

(4) After the acid elution, pass water through.

(5) Each drop from the bottom of the column is to be collected on a separate stainless steed source tray, the first drop to fall after the active solution has been placed on the top of the column is designated as eluant drop No. 1 and collected on tray No. 1.

(6) Flow rates throughout the elution should not exceed 2 drops per 3 minutes and for preference should be about half this rate.

(7) The column should be left wet with the reservoir above the bed filled with water and the column top sealed with blanked off glass joint.

(c) Analysis of Results

(1) Count each source tray independently for α and β activity.

(2) Plot the α count rate and the β count rate against drop numbers and indicate on the graph the positions where the eluant was changed.

Draw your conclusions regarding the efficiency of the method as a means of separation, the nature of the ions in solution under the conditions of the experiment, and the usefulness of tracers in the experiment.

Appendix—A
References

LABORATORY TECHNIQUES

Chase, G.D., "*Principles of Radioisotope Methodology,*" Burgess Publishing Co., Minneapolis, 1959.

Choppin, G.R., "*Experimental Nuclear Chemistry,*" Prentice-Hall, Inc., Englewood Cliffs, N.J., 1961.

Overman, R.T., and Clark, H.M., "*Radioisotope Techniques,*" McGraw-Hill Book Co., Inc., New York, 1960.

THEORETICAL MATERIAL

Evans, R.D., "*The Atomic Nucleus,*" McGraw-Hill Book Co., Inc., New York, 1955.

Friedlander, G., and Kennedy, J. W., "*Nuclear and Radio-chemistry,*" John Wiley & Sons, Inc., New York, 1955.

Gamow, G., "*Mr. Tompkins Explores the Atom,*" The University Press, Cambridge, England, 1944.

———"Mr. *Tompkins in Wonderland,*" The Macmillan Company, New York, 1945.

Glasstone, S., "*Sourcebook on Atomic Energy,*" 2nd ed., D. Van Nostrand Co., Inc., Princeton, N. J., 1958.

Halliday, D., "*Introductory Nuclear Physics,*" 2nd ed., John Wiley & Sons, Inc., New York, 1955.

Lapp, R.E., and Andrews, H.L., "*Nuclear Radiation Physics,*" 2nd 'ed., Prentice-Hall, Inc., Englewood Cliffs, N.J., 1954.

Price, W. J., "*Nuclear Radiation Detection,*" McGraw-Hill Book Co., Inc., New York, 1958.

Applications To Special Fields

Boyd, G.A., "*Autoradiography in Biology and Medicine,*" Academic Press, Inc., New York, 1955.

Comar, C.L., "*Radioisotopes in Biology and Agriculture,*" McGraw-Hill Book Co., Inc., New York, 1955.

Kohl, J., Zentner, R.D., and Lukens, H.R., "*Radioisotope Applications Engineering,*" D. Van Nostrand Co., Inc., Princeton, N. J., 1961.

U.S. Department of Health, Education and Welfare, Division of Radiological Health, "*Radiological Health Handbook,*" rev. ed., U.S. Dept. of Health, Education and Welfare, Washington, Sept., 1960 (PB-121784R).

Wahl, A.C., and Bonner, N.A., "*Radioactivity Applied to Chemistry,*" John Wiley & Sons, Inc. New York, 1951.

Appendix—B

Table 1

Characteristics of Relevant Fundamental Particles

(i) Stable Subatomic Particles

Particle	*Electron*	*Proton*	*Neutron*
Symbol	e	*p*	*n*
Location	outer sphere	nucleus	nucleus
Charge/e	–1	+1	0
Mass/u	0.000 55	1.007 277	1.008 665
Rest mass energy/Mev	0.511	938.259	939.552
Spin/h	1/2	1/2	1/2
Mean life/s	stable	stable	932
Decay products	—	—	$p+e^{-}+\overline{\nu}$
Antiparticle*	positron (e+)	antiproton (p^{-})	antineutron ($\overline{n}$)

(ii) Stable but appear only in some nuclear reactions

Particle	*Neutrino*	*Photon*
Symbol	ν	γ
Charge/e	0	0
Mass	0	0
Spin/h	½	1
Antiparticle*	antineutrino ($\overline{\nu}$)	self**

(iii) Unstable mesons

Particle	*Pion charged*	*Pion neutral*	*Muon*
Symbol	$\pi^{\pm}$	π^{*}	$\mu^{\pm}$
Charge	±1	0	±1
Mass/m_e	273	264	207
Rest mass energy/Mev	139.5	135	105.8
Spin/h	0	0	½
Mean life/s	2.6×10^{-8}	0.76×10^{-16}	2.2×10^{-6}
Decay products	$\mu^{\pm} + \nu$	2γ	$e^{\pm} + \nu$ or $\overline{\nu}$
Antiparticle*	$\pi^{\mp}$	self**	$\mu^{\mp}$

* An antiparticle differs from corresponding particle only in charge or spin orientation.

** Where a particle and its antiparticle cannot be distinguished.

Table 2

Properties of Nuclear Radiations

Radiation	*Identity*	*Energy spectrum*	*Penetration*	*Mechanism of energy loss*	*Energy spectra determination*
α (alpha)	He^{2+}	Discrete energies 4 to 9 MeV	3.5–11 mg/cm^2	Ionisation	Range in air rough), detector and pulse analyser
β (beta)	Electron	Continuous energy distbn. to E_{max} majority 0.5 to 2 MeV, E_{max}	200–1000 mg/cm^2	Ionisation Bremsstrahlung	Detector, pulse analyser or, for E_{max} only, aluminium absorber methods
γ (gamma)	Electro-magnetic radiation	Discrete energies Majority +0.1 to 2 MeV	$I = I_0 e^{-\mu x}$ $\mu = \frac{0.693}{x_{1/2}}$ $x_{1/2}$ for 1 MeV in Pb = ~ 1 cm	Photoeletctric Compton Pair production	Sodium iodide scintillation spectrometer

Mossbauer effect. Emission of γ-radiation from a nuclide and subsequent resonant absorption by same nucelar species by reduction in recoil energy. Spectrum scanned by relative motion of source and absorber to produce Doppler effect. Application to selected inorganic analysis and structure determination.

Szilard–Chalmers reaction. Rupture of chemical bonds by γ recoil energy from an n, γ reaction as a means of isolating an active nuclide from an isotopic target material.

Some Definitions

Roentgen	— That quantity of X- or γ-radiation which produces, as a consequence of ionisation, 1 e.s.u. of charge of either sign, in 1 cc of air measured at S.T.P.
Rad	— Unit of absorbed dose of any kind of radiation which is accompanied by liberation of 100 ergs of energy per gram of absorbing material.
Linear energy transfer (LET)	— is linear rate at which energy is dissipated along a particle or photon track (keV/μ).
G value	— Number of molecules changed for each 100 eV of energy absorbed.
Methods of radiation measurement	— Ion chamber (ionisation produced). Calorimetry (heat generated). Chemical dosimetry (chemical effect).

Chemical effects :

Table 3
Some uses and Applications of Isotopes

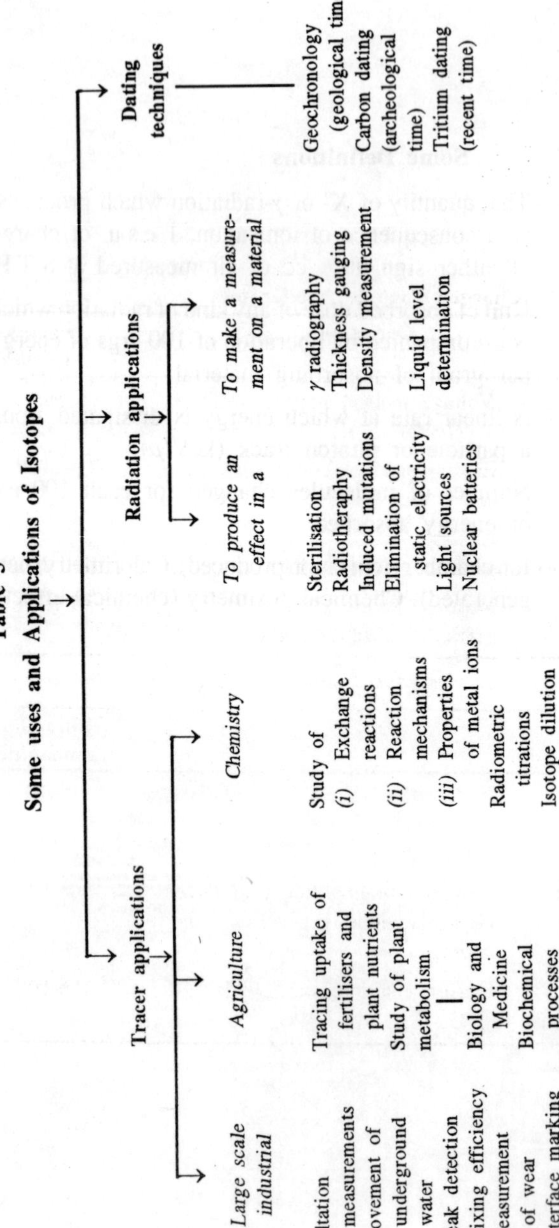

Table 4
Charged Particle Accelerators

Type	*Operation*	*Energy available*
Cockroft–Walton (1931)	Voltage multiplier used with accelerator tube.	1 MeV
Van de Graaff (1931)	Electrostatic generator used with accelerator tube.	10 MeV
Linear accelerator (1931) (Lawerence)	Multiple acceleration of charged particles in linear path.	1 BeV (1000 MeV)
Cyclotron (1931) (Lawerence)	Multiple acceleration of charged particles in spiral path.	20 MeV

Neutron Sources

Type	*Operation*	*Neutron flux available*
Radioactive sources	Radioactive nuclides to produce α, n and γ, n reactions with beryllium.	10^4–10^7
Particle accelerators	Deuteron bombardment of light elements, notably beryllium and tritium.	10^8–10^{10}
Nuclear reactor	Fission chain reaction with $^{235}_{92}U$, $^{233}_{92}U$ or $^{239}_{94}Pu$.	10^8–10^{15}

Table 5
Actinide Elements

Atomic Number	*Element*	*Valencies exhibited in aqueous solution*			*Notes*
		Stable	*Semi-stable*	*Unstable*	
89	Actinium (Ac)	3			
90	Thorium (Th)	4			
91	Protactinium (Pa)	5		4	
92	Uranium (U)	6	4	3,5	
93	Neptunium (Np)	6	4,5	3	
94	Plutonium (Pu)	6	3,4	5	
95	Americium (Am)	3		4,5,6	
96	Curium (Cm)	3			
97	Berkelium (Bk)	3	4		
98	Californium (Cf)	3			
99	Einsteinium (Es)	3			
100	Fermium (Fm)	3			
101	Mendelevium (Md)	3			
102	Nobelium (No)	3			
103	Lawrencium (Lw)	3			

Table 6

Nuclear Reactors in India

Reactor	*Type*	*Neutron flux/s*	*Power /MW*	*Fuel*	*Cladd-ing*	*Mode-rator*	*Cooland prim/sec*	*Purpose*	*Date of criticality*
Apsara	Swimming pool	1.25×10^{13}	1.0	U*	Al	lw	lw	Research and isotope prod.	1956 Aug.
Cirus	tank	6.7×10^{13}	40	U	Al	hw	hw/lw	Isotope prod.	1960 July
Zerlina	tank		0	U	Al	hw	hw/lw	Research	1961 July
Purnima I	fast		0	U-Pu oxide	St. steel	—	air	Research	1972 May
Tarapur	boiling water		2×200	U*	Zr-Al	lw	lw	Power gen.	1969
Kota	PHWR**		2×220	U	Zr-Al	hw	hw/lw	Power gen.	1973
Purnima II	fast		0	U-Pu oxide	St. steel	—	air	Research	1976
Kalpakkam I	PHWR	1.3×10^{14}	2×235	U	Zr-Al	hw	hw/lw	Power gen.	1983
Kalpakkam Ii	PHWR			U	Zr-Al	hw	hw/lw	Power gen.	1985 Aug.
Dhruva	tank	1.3×10^{14}	100	U	Al	hw	hw/lw	Research and isotope prod.	1985 Aug.
Fast breeder	loop		13	PuC (70) +UC(30)	St. Steel	—	liquid sodium	Research	1985 Dec.

**PHWR : Pressurised heavy water under helium pressure. *U : Enriched uranium